David L. Andrews · Lasers in Chemistry

Springer

Berlin
Heidelberg
New York
Barcelona
Budapest
Hong Kong
London
Milan
Paris
Santa Clara
Singapore
Tokyo

David L. Andrews

Lasers in Chemistry

Third Edition

With 121 Figures

 Springer

Professor David L. Andrews

School of Chemical Sciences
University of East Anglia
Norwich NR4 7TJ
U.K.

ISBN-3-540-61982-8   3. Aufl. Springer-Verlag Berlin Heidelberg New York
ISBN-3-540-51777-4   2. Aufl. Springer-Verlag Berlin Heidelberg New York

Die Deutsche Bibliothek – CIP-Einheitsaufnahme
Andrews, David L.:
Lasers in chemistry / David L. Andrews. – 3. ed. – Berlin ; Heidelberg ;
New York ; Barcelona ; Budapest ; Hong Kong ; London ; Milan ; Paris ;
Stanta Clara ; Singapore ; Tokyo : Springer, 1997
    ISBN 3-540-61982-8 kart.

Springer-Verlag Berlin Heidelberg 1990, 1997
Printed in Germany

Typesetting: MEDIO, Berlin
Cover design: de'blik, Berlin
SPIN: 10515405   52/3020 – 5 4 3 2 1 0 – Printed on acid-free paper

To my wife Karen,
whose love and encouragement are inexpressibly precious

Preface to the Third Edition

Ten years have now elapsed since this book first came into print – time enough in this fast-moving field to warrant preparation of a new, third edition. In a host of minor but significant technical details the new edition has been brought up to date, but also there are many more substantial additions reflecting new technology, methods and applications. Through a wide range of innovations – notably with the titanium:sapphire laser, new beam profile technology, and charge-coupled device instrumentation – many areas that were once the province of the lonely few have acquired a host of new practitioners and important new applications. Laser mass spectrometry and, in particular, the technique of matrix-assisted laser desorption ionisation, is one obviously burgeoning field; so, too, is the study of protein and other biological structures using Raman optical activity, with materials processing also deserving mention. With many other already well-established techniques there have been significant developments not only through increasing applications but also through better understanding of the principles at work, as for example with fluorescence energy migration and photodynamic therapy, the latter now substantially less of a chemical and biological mystery than it was only a few years ago.

My particular thanks are due to the following whose expertise is reflected in some of the new material; Professor Laurence Barron, Mike Cudby, Andrey Demidov, David James, David Russell, and Professor Norman Sheppard. Finally, I must thank my colleague Steve Meech, to whose experimental acumen I owe my own involvement with the femtosecond laser technology heralded in the Preface to the Second Edition.

Norwich, June 1997   David L. Andrews

Preface to the Second Edition

In the three and a half years which have passed since publication of the First Edition, the pace of development in laser technology has continued to further the extent to which lasers are applied in Chemistry. Laser photochemical techniques are increasingly being employed in fundamental chemical research, and many analytical methods such as laser mass spectrometry are moving across the boundary which divides academic research from routine instrumental analysis. This change is now being reflected in the introduction of teaching on laser methods in many, if not most Chemistry degree courses.

A number of important innovations have also arisen since the appearance of the First Edition, and a few of those represented in the new edition are worth singling out for mention. One of the most significant has undoubtedly been the application of ultrafast spectroscopy based on new femtosecond pulsed lasers. The measurement of a timescale for chemical bond breaking is a particularly nice illustration of the hitherto inaccessible information derivable from such studies. From the more pragmatic viewpoint of the analyst, perhaps more significant has been the arrival of Fourier transform Raman spectrometry, a technique which is already having a revolutionary impact on vibrational spectroscopy. Other synthetic chemists are showing increasing interest in the design and characterisation of novel materials for nonlinear optical applications. The interplay between lasers and chemistry is one which provides the clearest illustration of the phenomenally interdisciplinary nature of modern optics-based research.

In preparing a Second Edition I have endeavoured as far as possible to retain a balance between coverage of the more widely used laser methods and new developments which at present still lie firmly in the province of academia. I have also largely preserved the original structure in terms of Chapters and Sections. The principal addition is the inclusion for the first time of a Questions Section at the end of each Chapter. Answers to the numerical problems are also included at the end of the book. I hope this will help serve the needs of those who have adopted this as a set course book.

I am indebted to many people for their encouraging and helpful comments on the First Edition, most especially Professor Jim Turner and Martin McCoustra. I am also pleased to record my indebtedness to Rainer Stumpe of Springer-Verlag for his interest and patience in waiting for this Second Edition.

Norwich, January 1990   David L. Andrews

Preface

Let us try as much as we can,
we shall still unavoidably fail in many things
'The Imitation of Christ', Thomas a Kempis

Since the invention of the laser in 1960, a steadily increasing number of applications has been found for this remarkable device. At first it appeared strangely difficult to find any obvious applications, and for several years the laser was often referred to as 'a solution in search of a problem'. The unusual properties of laser light were all too obvious, and yet it was not clear how they could be put to practical use. More and more applications were discovered as the years passed, however, and this attitude slowly changed until by the end of the 1970's there was scarcely an area of science and technology in which lasers had not been found application for one purpose or another. Today, lasers are utilised for such diverse purposes as aiming missiles and for eye surgery; for monitoring pollution and for checking out goods at supermarkets; for welding and for light-show entertainment. Even within the field of specifically chemical applications, the range extends from the detection of atoms at one end of the scale to the synthesis of vitamin D at the other.

In this book, we shall be looking at the impact which the laser has made in the field of chemistry. In this subject alone, the number of laser applications has now grown to the extent that a book of this size can provide no more than a brief introduction. Indeed, a number of the topics which are here consigned to a brief sub-section are themselves the subjects of entire volumes by other authors. However, this itself reflects one of the principal motives underlying the writing of this book; much of the subject matter is elsewhere only available in the form of highly sophisticated research-level treatments which assume that the reader is thoroughly well acquainted with the field. It is the intention in this volume to provide a more concise overview of the subject accessible to a more general readership, and whilst the emphasis is placed on chemical topics, most of the sciences are represented to some extent. The level of background knowledge assumed here corresponds roughly to the material covered in a first-year undergraduate course in chemistry. There has been a deliberate effort to avoid heavy mathematics, and even the treatment of the fundamental laser theory in Chap. 1 has been simplified as much as possible.

The link between lasers and chemistry is essentially a three-fold one. Firstly, several important chemical principles are involved in the operation of most lasers. This theme is explored in the introductory Chap. 1 and 2; Chap. 1 deals with the chemical and physical principles of laser operation in general terms, and Chap. 2 provides a more detailed description of specific laser sources. Secondly, a large number of techniques based on laser instrumentation are used to probe systems of chemical interest. The systems may be either chemically stable or in the process of chemical reaction, but in each case the laser can be used as a highly sensitive analytical device. There are numerous ways in which laser instrumentation is now put to use in chemical laboratories, and some of the general principles are discussed in Chap. 3. However, by far the most widely used methods are specifically spectroscopic in nature, and Chap. 4 is therefore entirely given over to a discussion of some of the enormously varied chemical applications of laser spectroscopy.

The third link between lasers and chemistry concerns the inducing of chemical change in a system through its irradiation with laser light. In such applications, the chemistry in the presence of the laser radiation is often quite different from that observed under other conditions; thus, far from acting as an analytical probe, the radiation is actively involved in the reaction dynamics, and acts as a stimulus for chemical change. This area, discussed in Chap. 5, is one of the most exciting and rapidly developing areas of chemistry, and one in which it appears likely that the laser may ultimately make more of an impact than any other.

An attempt has been made throughout this book to provide examples illustrating the diversity of laser applications in chemistry across the breadth of the scientific spectrum from fundamental research to routine chemical analysis. Nonetheless the emphasis is mostly placed on applications which have relevance to chemical industry. The enormous wealth of material from which these illustrations have been drawn means that this author's choice is inevitably idiosyncratic, although each example is intended to provide further insight into the underlying principles involved. Since this is an introductory textbook, references to the original literature have been kept to a minimum in order to avoid swamping the reader; needless to say, this means that a great many of the pioneers of the subject are not represented at all. I can only record my immense debt to them and all who have contributed to the development of the subject to the state in which I have reported it.

I should like to express my thanks to the editorial staff of Springer-Verlag, and especially Almut Heinrich who has been most sympathetic and helpful at all times. Thanks are also due to the staff of Technical Graphics for their work on the original Figures. I am indebted to all who have contributed spectra and consented to the reproduction of diagrams, and I am very grateful to Peter Belton, Godfrey Beddard, Nick Blake, John Boulton, Colin Creaser, Allan Dye, U. Jayasooriya and Ron Self for comments on various parts of the manuscript. Lastly I should like to express my thanks to my good friend and severest critic John Sodeau, whose comments have done more than any other to help remould my first draft into its completed form.

Norwich, June 1986   David L. Andrews

Table of Contents

CHAPTER 1

Principles of Laser Operation

*And the atoms that compose this radiance do not travel as isolated
individuals but linked and massed together*
'De Rerum Naturae', Lucretius

1.1
The Nature of Stimulated Emission

The term *laser*, an acronym for *l*ight *a*mplification by the *s*timulated *e*mission
of *r*adiation, first appeared in 1960 and is generally held to have been coined
by Gordon Gould, one of the early pioneers of laser development. Since the
device was based on the same principles as the *maser*, a microwave source
which had been developed in the 1950s, the term 'optical maser' was also in
usage for a time, but was rapidly replaced by the simpler term. In order to
appreciate the concepts of laser action, we need to develop an understanding
of the important term 'stimulated emission'. First, however, it will be helpful to
recap on the basic quantum mechanical principles associated with the absorption and emission of light. Although these principles apply equally to individual ions, atoms or molecules, it will save unnecessary repetition in the following discussion if we simply refer to molecules.

According to quantum theory, molecules possess sets of discrete energy levels, and the energy which any individual molecule can possess is limited to
one of these values. Broadly speaking, the majority of molecules at any moment in time exist in the state of lowest energy, called the ground state, and it
is often by the absorption of light that transitions to states of higher energy
can take place. Light itself consists of discrete quanta known as photons, and
the absorption process thus occurs as individual photons are intercepted by
individual molecules; in each instance the photon is annihilated and its energy
transferred to the molecule, which is thereby promoted to an excited state. For
this process to occur the photon energy E, which is proportional to its frequency ν, (E $=$ hν, where h is Planck's constant,) must match the gap in energy
between the initial and final state of the molecule. Since only discrete energy
levels exist, there results a certain selectivity over the frequencies of light that
can be absorbed by any particular compound; this is the principle underlying
most of spectroscopy.

What we now need to consider in more detail is the reverse process, namely the emission of light. In this case we start off with a system of molecules, many of which must already exist in some excited state. Although the means for producing the excitation is not important, it is worth noting that it need not necessarily be the absorption of light; for example, in a candle it is the chemical energy of combustion which provides the excitation. Molecules in excited states generally have very short lifetimes (often between 10^{-7}s and 10^{-11}s) and by releasing energy they rapidly undergo relaxation processes. In this way, the molecules undergo transitions to more stable states of lower energy, frequently the ground state. There are many different mechanisms for the release of energy, some of which are radiative, in the sense that light is emitted, and some of which are non-radiative. However, although chemical distinctions can be made between different types of radiative decay, such as fluorescence and phosphorescence, the essential physics is precisely the same – photons are emitted which precisely match the energy difference between the excited state and the final state involved in the transition. Because this kind of photon emission can occur without any external stimulus, the process is referred to as *spontaneous emission.*

We are now in a position to come to terms with the very different nature of stimulated emission. Suppose, once again, we have a system of molecules, some of which are in an excited state. This time, however, a beam of light is directed into the system, with a frequency such that the photon energy exactly matches the gap between the excited state and some state of lower energy. In this case, each molecule can relax by emitting another photon of the same frequency as the supplied radiation. However, it turns out that the probability of emission is enhanced if other similar photons are already present. Moreover, emission occurs preferentially in the direction of the applied beam, which is thereby amplified in intensity. This behaviour contrasts markedly with the completely random directions over which spontaneous emission occurs when no beam is present. This type of emission is therefore known as 'stimulated emission': it is emission which is stimulated by other photons of the appropriate frequency. The two types of emission are illustrated in Fig. 1.1.

The concept of stimulated emission was originally developed by Einstein in a paper dealing with the radiation from a heated black body, a subject for which it had already been found necessary to invoke certain quantum ideas (Planck's hypothesis) in order to obtain a theory which fitted experimental data. Despite publication of this work in 1917, and subsequent confirmation of the result by the more comprehensive quantum mechanics in the late 1920s, little attention was paid to the process by experimental physicists for many years. It was only in the early 1950s that the first experimental results appeared, and the practical applications began to be considered by Gould, Townes, Basov and Prochorov, and others. The key issue in the question of applications was the possibility of amplification implicit in the nature of the stimulated emission process. These considerations eventually led to develop-

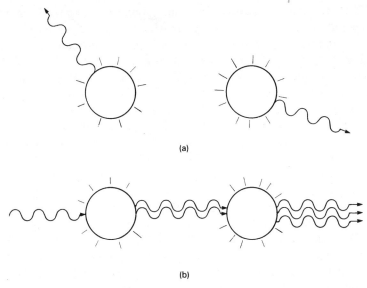

Fig. 1.1a and b Spontaneous (a) and stimulated (b) emission by excited molecules

ment of the maser, the forerunner of the laser which produced microwave rather than optical frequency radiation. It was undoubtedly the success in producing an operational device based on stimulated microwave emission that spurred others on towards the goal of a visible light analogue.

1.2
Resonators and Pumping Processes

We can now consider some of the fundamental practical issues involved in the construction of a laser. For the present, we shall consider only those features which directly follow from the nature of the stimulated emission process on which the device is based; other considerations will become evident later in this chapter, and a full account of particular laser systems is given in the following chapter. At this stage, however, it is clear that the first requirement is a suitable substance in which stimulated emission can take place, in other words an *active medium*. We also require an external stimulus in order to promote atoms or molecules of this medium to an appropriate excited state from which emission can occur.

The active medium can take many forms, gas, liquid or solid, and the substance used is determined by the type of output required. Each substance has its own unique set of permitted energy levels, so that the frequency of light emitted depends on which levels are excited, and how far these are separated in energy from states of lower energy. The first laser, made by Maiman in 1960, had as its active medium a rod of ruby and produced a deep red beam of light. More common today are the gas lasers, in which gases such as argon or carbon dioxide form the active medium; the former system emits various frequencies of visible light, the latter infra-red radiation.

There is an equal diversity in the means by which the initial excitation may be created in the active medium. In the case of the ruby laser, a broadband source of light such as a flashlamp is used; in gas lasers an electrical discharge provides the stimulus. Later, we shall see that even chemical reactions can provide the necessary input of energy in certain types of laser. Two general points should be made, however, concerning the external supply of energy. First, if electromagnetic radiation is used, then the frequency or range of frequencies supplied must be such that the photons which excite the laser medium have an energy equal to or larger than that of the laser output. Secondly, because of heat and other losses no laser has 100% efficiency, and the energy output is always less than the energy input in the same way as in any electronic amplifier. These rules are simply the results of energy conservation. One other point is worth mentioning at this juncture. So far, we have implicitly assumed that the excitation precedes laser emission; indeed this is true for many kinds of laser system. However, for a continuous output, we are faced with a requirement to sustain a population of the generally short-lived excited molecules. We shall see shortly that this imposes other requirements on the nature of the active medium.

As it stands, we have described a set-up in which a laser medium can become excited and relax by the emission of photons. We do not have to supply another beam of light for stimulated emission to occur; once a few molecules have emitted light by spontaneous emission, these emitted photons can stimulate emission from other excited molecules. However, a single passage of photons through the medium is not generally sufficient for stimulated emission to play a very significant role. This is because the rate of stimulated emission is proportional to the initial intensity, which as we have pictured the process so far cannot be very great.

For this reason, it is generally necessary to arrange for the multiple passage of light back and forth through the active medium in a *resonator*, so that with each traversal the intensity can be increased by further stimulated emission. In practice this can be arranged by placing parallel mirrors at either end of the laser medium, so that light emitted along the axis perpendicular to these mirrors is essentially trapped and bounces backwards and forwards indefinitely, growing in intensity all the time. By contrast, photons which are spontaneously emitted in other directions pass out of the active medium and no longer contribute to stimulated emission. Figure 1.2 illustrates how the intensity of light travelling between the mirrors builds up at the expense of spontaneous emission over a short period of time; since emission along the laser axis is self-enhancing, we very rapidly reach the point at which nearly all photons are emitted in this direction.

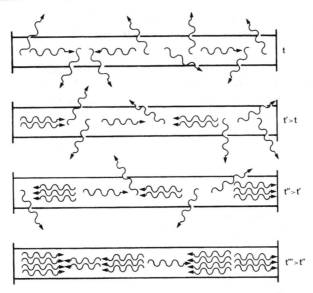

Fig. 1.2 The development of laser action from the onset of photon emission at time t through three later instants of time

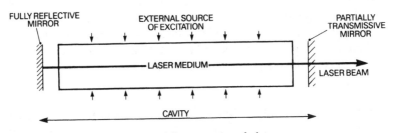

Fig. 1.3 Essential components of a laser

We have seen in this section that the resonator plays an important role in the operation of any laser. To conclude, we briefly mention two other practical considerations. Although we have talked in terms of two parallel plane end-mirrors, the difficulty of ensuring sufficiently precise planarity of the mirrors, coupled with difficulty of correct alignment, has led to several other configurations being adopted. One of the most common is to have two confocal concave mirrors, which allow slightly off-axis beams to propagate back and forth. The other crucial consideration is, of course, the means of obtaining usable output of laser light from the resonator beam; this is generally accomplished by having one of the end-mirrors made partially transmissive. Figure 1.3 schematically illustrates the basic components of the laser as discussed so far.

1.3
Coherent Radiation, Standing Waves and Modes

Stimulated emission, as we have seen, results in the creation of photons with the same frequency and direction as those which induce the process. In fact, the stimulated photons are identical to the incident photons in every respect, which means that they also have the same phase and polarisation; this much is implied in Fig. 1.2. It is the fact that the inducing and emitted photons have a phase relationship that lies behind use of the term *coherent* to describe laser radiation. In the quantum theory of light, it is usual to refer to *modes* of radiation, which simply represent the possible combinations of frequency, polarisation and direction which characterise photons. In the laser, only certain modes are permitted, and these are the modes which can result in the formation of standing waves between the end-mirrors. This means that these mirrors, together with everything inbetween them, including the active medium, must form a *resonant cavity* for the laser radiation; it is for this reason that the term 'cavity' is often used for the inside of a laser.

The standing wave condition means that nodes, in other words points of zero amplitude oscillation, must be formed at the ends of the laser cavity. This condition arises from the coherent nature of the stimulated emission process; since there is a definite phase for the radiation, then it is necessary to ensure that only constructive interference, and not destructive interference, occurs as light travels back and forth inside the cavity. The only way to satisfy the standing wave condition is if it is possible to fit an integer number of half-wavelengths into the cavity, as illustrated in Fig. 1.4. Hence the wavelengths λ which can resonate inside the laser cavity are determined by the relation

$$m\lambda/2 = L, \tag{1.1}$$

where m is an integer, and L is the cavity length. Standing waves, of course, crop up in many different areas of physics; the vibrations of a violin string, for example, are subject to the same kind of condition. However, in the laser we are typically dealing with several million half-wavelengths in the cavity, and the range of wavelengths emitted is constrained by the molecular transitions of the active medium, not just the standing wave condition; we shall examine this more fully in Sect. 1.6.4.

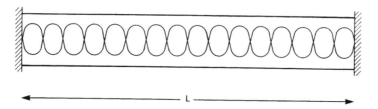

Fig. 1.4 A standing wave of electromagnetic radiation between two mirrors

Although we have been discussing a standing wave based on propagation of radiation back and forth exactly parallel to the laser axis, as shown in Fig. 1.5a, lasers can also sustain resonant oscillations involving propagation slightly off-axis, as shown in Fig. 1.5.b. Such modes are distinguished by the label TEM_{pq}, in which the initials stand for 'transverse electromagnetic mode', and the subscripts p and q take integer values determined by the number of intensity minima across the laser beam in two perpendicular directions. Figure 1.5c illustrates the intensity patterns for three simple cases. Although it may be possible to increase the intensity of laser output by allowing *multi-mode* operation, it is generally more desirable to suppress all but the axial TEM_{00} mode, known as the *uniphase* mode, which has better coherence properties. Uniphase operation is generally ensured by adopting a narrow laser cavity.

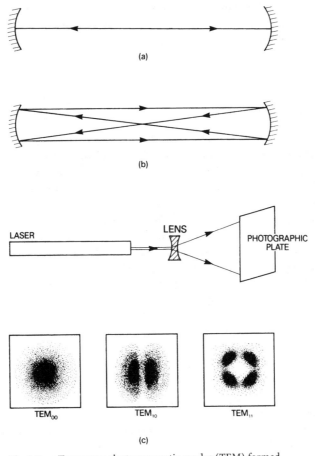

Fig. 1.5a–c Transverse electromagnetic modes (TEM) formed with confocal concave mirrors; **(a)** TEM_{00}, **(b)** TEM_{10}. In **(c)**, the distribution of intensity in the laser beam cross-section is illustrated for three different modes

1.4
The Kinetics of Laser Emission

1.4.1
Rate Equations

We are now in a position to take a look at some crucial theoretical constraints on laser operation. The simplest approach to a more detailed understanding of laser dynamics is to take a model system in which there are only two energy levels E_1 and E_2 involved; to start with, let us also assume that both are non-degenerate. There are, as we discussed earlier, three distinct radiative processes that can accompany transitions between the two levels: absorption, spontaneous emission, and stimulated emission, illustrated in Fig. 1.6. For simplicity, we shall for the present ignore non-radiative processes. The transition rates corresponding to the three radiative transitions, due to Einstein, may be written as follows:

$$R_{absorption} = N_1 \rho_\nu B_{12}; \tag{1.2}$$

$$R_{spont.emission} = N_2 A_{21}; \tag{1.3}$$

$$R_{stim.emission} = N_2 \rho_\nu B_{21}. \tag{1.4}$$

Here, N_1 and N_2 are the numbers of molecules in each of the two energy levels E_1 and E_2 respectively, ρ_ν is the energy density of radiation with frequency ν, and A_{21}, B_{21} and B_{12} are known as the Einstein coefficients. (Note that originally the Einstein coefficients were defined as the constants of proportionality in rate equations for polychromatic radiation and were thus expressed in different units in terms of the energy density *per unit frequency interval*. However, the definition used here is gaining acceptance since it is more appropriate for application to essentially monochromatic laser radiation.) For a non-degenerate pair of energy levels, it is readily shown that the two B coefficients are equal, and we can therefore drop the subscripts. It also transpires that the A coefficient has a cubic dependence on frequency, so that the higher the excited state energy, the more its decay is dominated by spontaneous emission. This is why it is significantly more difficult to achieve laser action at short (particularly UV and X-ray) wavelengths.

Now we need to consider the intensity of light travelling along the laser axis. For a parallel beam, this is best expressed in terms of the *irradiance* I, defined as the energy crossing unit cross-sectional area per unit time. This can be directly related to the instantaneous photon density, as follows. Consider a small cube of space of side length l, and volume V, through which the beam passes. If at any instant this cube contains N photons, the photon density ϕ is N/V, and the cube contains an energy $Nh\nu$. Now if the laser medium has refractive index n, then since each photon takes a time l/c' to traverse the cube, where $c' = c/n$ is the velocity of light within the medium, it clearly takes the

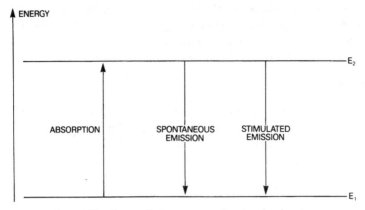

Fig. 1.6 The three radiative processes in a two-level system

same time l/c' for the energy $Nh\nu$ to pass through the cube. Hence the irradiance is given by

$$I = \frac{\text{energy per unit time}}{\text{area}} = \frac{Nh\nu(1/c')}{1^2}, \tag{1.5}$$

and since $1^3 = V$, we obtain the following relation for the instantaneous photon density:

$$\phi = N/V = I/hc'\nu. \tag{1.6}$$

We can also relate this to the energy density ρ_ν, since this is simply $h\nu$ times the above result:

$$\rho_\nu = I/c'. \tag{1.7}$$

Returning to the rate Eqs. (1.2) to (1.4), we note that since spontaneous emission occurs in random directions, it cannot appreciably contribute to an increase in the intensity of light propagating along the laser axis, since very few photons will be spontaneously emitted in this direction. Therefore if we write down an equation for the rate of increase of intensity along the laser axis, we have only two major terms, namely a positive term due to stimulated emission, and a negative term due to absorption. Using the results (1.6) and (1.7), we thus have the following result for the rate of change in photon density:

$$d\phi/dt = (hc'\nu)^{-1}dI/dt \approx N_2B(I/c') - N_1B(I/c'), \tag{1.8}$$

and hence

$$dI/dt \approx (N_2 - N_1)BIh\nu. \tag{1.9}$$

Strictly speaking, there are other terms, such as the intensity loss due to imperfect reflectivity of the end-mirrors, which ought to be included in Eq. (1.9), but for the moment we shall ignore them.

What is now apparent is that the rate of increase of intensity is proportional to the difference in the excited state and ground state populations, $(N_2 - N_1)$. Clearly for the amplification we require, we need to have a positive value for dI/dt, which leads to the important condition for laser action:

$$dI/dt > 0 \Rightarrow N_2 > N_1. \tag{1.10}$$

Physically, this means that unless we have more molecules in the excited state than the ground state, absorption will win over stimulated emission. If we extend our theory to include the effects of degeneracy in the two energy levels, the counterparts to Eqs. (1.9) and (1.10) are:

$$dI/dt \approx (N_2 - (g_2/g_1)N_1)BIh\nu, \tag{1.11}$$
$$dI/dt > 0 \Rightarrow N_2 > N_1(g_2/g_1), \tag{1.12}$$

where g_1 is the degeneracy of level 1, and g_2; that of level 2. The interesting thing about the latter inequality is that it can never be satisfied under equilibrium conditions. For a system in thermal equilibrium at absolute temperature T, the ratio of populations is given by the Boltzmann relation;

$$N_2/N_1 = (g_2/g_1)\exp((E_1 - E_2)/kT), \tag{1.13}$$

where k is Boltzmann's constant. Since $E_2 > E_1$, it follows that (1.13) can only hold for the physically meaningless case of a negative absolute temperature! The result illustrates the fact that laser operation cannot occur under the equilibrium conditions for which the concept of temperature alone has meaning. Indeed, at thermal equilibrium, the converse of Eq. (1.12) always holds true – in other words, the population of the state with lower energy is always largest. The situation we require in a laser is therefore often referred to as a *population inversion*, and energy must be supplied to the system to sustain it in this state. The means by which energy is provided to the system is generally known as *pumping*.

The solution to Eq. (1.11) may be written in the form

$$I = I_0 \exp(at), \tag{1.14}$$

where

$$a = (N_2 - (g_2/g_1)N_1)Bh\nu. \tag{1.15}$$

Alternatively, the intensity may be expressed in terms of the distance x travelled by the beam, since $t = x/c'$. Hence we may write

$$I = I_0 \exp(kx), \tag{1.16}$$

where k is known as the small signal gain coefficient, and is defined by

$$k = (N_2 - (g_2/g_1)N_1)Bh\nu/c'. \tag{1.17}$$

1.4.2
Threshold Conditions

We can now make our theory a little more realistic by considering the question of losses. Quite apart from the imperfect reflectivity of the end-mirrors, there must always be some loss of intensity within the laser cavity due to absorption, non-radiative decay, scattering and diffraction processes. In practice, then, even with perfectly reflective mirrors the distance-dependence of the beam should more correctly be represented by the formula

$$I = I_0 \exp((k - \gamma)x), \tag{1.18}$$

where γ represents the net effect of all such loss mechanisms. However, since one end-mirror must be partially transmissive in order to release the radiation, and the other will not in practice be perfectly reflective, then in the course of a complete round-trip back and forth along the cavity (distance 2L, where L is the cavity length) the intensity is modified by a factor

$$G = I/I_0 = R_1 R_2 \exp(2L(k - \gamma)), \tag{1.19}$$

where R_1 and R_2 are the reflectivities of the two mirrors. The parameter G is known as the *gain*, and clearly for genuine laser amplification we require $G > 1$. For this reason, there is always a certain *threshold* for laser action; if $G < 1$, the laser acts much more like a conventional light source, with spontaneous emission responsible for most of the output. The threshold value of the small signal gain coefficient k is found by equating the right-hand side of Eq. (1.19) to unity, giving the result:

$$k_{threshold} = \gamma + (2L)^{-1} \ln(1/R_1 R_2). \tag{1.20}$$

For a laser with known loss and mirror reflectivity characteristics, the right-hand side of Eq. (1.20) thus represents the figure which must be exceeded by the gain coefficient, as defined by Eq. (1.17), if laser action is to occur. This in turn places very stringent conditions on the extent of population inversion required.

1.4.3
Pulsed Versus Continuous Emission

Thus far, the conditions for laser operation have been discussed without considering the possibility of pulsed emission. However, we shall discover that whilst some lasers operate on a *continuous-wave* (cw) basis, others are inherently pulsed. In order to understand the reason for this difference in behaviour, we have to further consider the kinetics of the pumping and emission processes and the physical characteristics of the cavity. To start with, we make the obvious remark that if the rate of pumping is exceeded by that of decay from the upper laser level, then a population inversion cannot be sustained, and pulsed operation must ensue, with a pulse duration governed by the decay kinetics. For continuous operation, we therefore require a pumping mechanism that is tailored to the laser medium.

Next, we note that the round-trip time for photons travelling back and forth between two parallel end-mirrors is given by

$$\tau = 2L/c'. \tag{1.21}$$

If the two end-mirrors were perfectly reflective, the number of round trips would be infinite, if we ignore absorption and off-axis losses. However, with end-mirror reflectivities R_1 and R_2, the probability of photon retention during any one round-trip within the laser cavity is R_1R_2, the probability of escape is $(1 - R_1R_2)$, and the mean number of round-trips per photon is therefore given by

$$z = (1 - R_1R_2)^{-1}. \tag{1.22}$$

Hence the mean time spent by a photon before leaving the cavity is

$$t = \frac{2L}{c'(1 - R_1R_2)}. \tag{1.23}$$

Continuous-wave operation is therefore possible provided the mean interval between the emission of successive photons by any single atom or molecule of the active medium is comparable to t.

In the next section, we shall leave behind the simple two-level model and consider more realistic energy level schemes on which lasers can be based. When there are more than two levels involved in laser action, there are also many more transitions participating, and the cycle which starts and ends with the active species in its ground state generally includes a number of excitation, emission and decay processes. Under such circumstances, the criterion for cw operation is that the timescale for a complete cycle of laser transitions must be similar to the mean cavity occupation time. We also note that whilst the converse is not feasible, it is possible to modify continuous-wave lasers so as to produce pulsed output. This is accomplished by the inclusion of electro-optical devices within the cavity; the principles are discussed in detail in Sect. 3.3.

1.5
Transitions, Lifetimes and Linewidths

We have seen that the competition between absorption and stimulated emission provides a problem in a two-level system. Indeed if the initial excitation is optically induced it is a problem which cannot be solved, for absorption will always win over stimulated emission. For a more workable proposition we therefore need to look at systems where more than two levels can be involved in laser action. In this section we begin by considering three- and four-level systems, both of which are widely used in lasers, before making some further general remarks concerning emission linewidths. Because of the greatly increased complexity of the rate equations when more than two levels are involved, a less mathematical approach is adopted for this section.

1.5.1
Three-Level Laser

In a three-level laser, two levels (E_1 and E_2) must, of course, be coupled by the downward transition which results in emission of laser light. Suppose then, that a third, and higher energy level E_3 is present, and that it is this level which is first populated by a suitable pumping mechanism. (As noted earlier, the energy of the emitted photon cannot exceed the energy supplied to excite a molecule of the laser medium – hence it has to be the state of highest energy that is initially populated in this case.) What we require for laser action is that a population inversion is established between levels E_1 and E_2, and this can be accomplished as follows.

With three different energy levels, there are three pairs of levels between which transitions can occur. It is the possibility of one of the downward transitions being a non-radiative process which makes all the difference; such radiationless processes can result from a number of different mechanisms such as bimolecular collisions or wall collisions in gases, or lattice interactions in the solid state. Thus if level E_3 rapidly undergoes radiationless decay down to level E_2, then, provided the laser emission is slower, the population of this level can increase while that of the ground state is diminishing because of pump transitions to E_3. Obviously, for the laser emission to be slow we require state E_2 to be metastable, with a comparatively long lifetime (say 10^{-3} s). Hence we can obtain the required population inversion, and laser action results.

Figure 1.7 illustrates the energy levels and populations for this arrangement. One difficulty is that lasing automatically repopulates the ground state, making

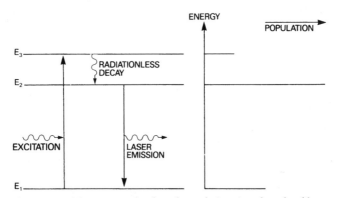

Fig. 1.7 Transitions, energy levels and populations in a three-level laser

it somewhat difficult to sustain the population inversion. In such a laser, once a population inversion is established all the atoms or molecules of the active medium may cascade down to the ground state together, resulting in the emission of a short sharp burst of light. The earliest ruby laser (see Sect. 2.1.1) was of this type, emitting pulses of about 1-ms duration, and requiring something like a minute between pulses to build up the population inversion again; this provides us with a good example of a laser that inherently operates in a pulsed mode.

1.5.2
Four-Level Laser

The problems associated with ground state repopulation may be overcome in a four-level laser. Here the energy level scheme is quite similar to the three-level case we have just considered; the only difference is that the lower of the two laser levels involved in the radiative transition lies above the ground state, as shown in Fig. 1.8. If this state rapidly decays to the ground state below, then it can never build up an appreciable population, and so population inversion *between the levels involved in lasing*, i.e. E_2 and E_3, is retained. Thus even though the ground state population may exceed that of any other level, laser action can still occur. In practice, this kind of scheme is quite common; a good example is the neodymium-YAG laser which we shall look at in more detail in the next chapter (Sect. 2.1.2). Since a population inversion can be sustained in this kind of laser, it can be made to operate in either a continuous or a pulsed mode.

1.5.3
Emission Linewidths

Ideally, each transition responsible for laser emission should occur at a single, well-defined frequency determined by the spacing between the energy levels it connects. However, there are several *line-broadening* processes which result in statistical deviations from the ideal frequency, so that in practice the emission has a frequency distribution like that shown in Fig. 1.9. The quantity normally referred to as the *linewidth*, or *halfwidth*, is a measure of the breadth of this distribution and is more accurately defined as the *full width at half maximum*, or FWHM for short. The various line-broadening mechanisms responsible for the frequency distribution can be divided into two classes. Those which apply equally to all the atoms or molecules in the active medium responsible for emission are known as *homogeneous* line-broadening processes and generally produce a Lorentzian frequency profile, and those based on the statistical differences between these atoms or molecules are known as *inhomogeneous* line-broadening processes, and generally result in a Gaussian profile. Usually a

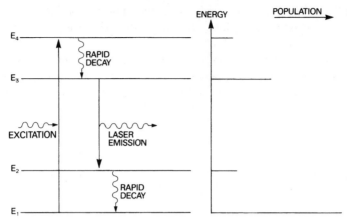

Fig. 1.8 Transitions, energy levels and populations in a four-level laser

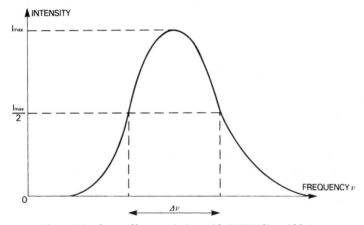

Fig. 1.9 Lineshape of laser emission, with FWHM linewidth $\Delta\nu$

number of these mechanisms operate simultaneously, each contributing to an increase in the emission linewidth.

The particulars of these processes vary according to the nature of the active medium. The one homogeneous mechanism which applies universally is that which arises from a frequency–time uncertainty relation. If laser photons are emitted from an excited state associated with a lifetime Δt, then there is a minimum uncertainty $\Delta\nu \geq (1/2\pi\Delta t)$ in the frequency of the photons emitted, resulting in emission line-broadening. This process is known as *natural line-broadening*. The lifetime of any excited state is, of course, influenced by the various decay pathways open to it, including radiative decay (both spontaneous and stimulated emission), radiationless decay and collisional energy transfer. The same frequency–time uncertainty relation also has an important bearing on the linewidth in lasers producing pulses of very short duration, as we shall see in Sect. 3.3.3.

The other major homogeneous line-broadening mechanisms are, briefly, as follows. In crystalline media, lattice vibrations cause a time-dependent variation in the positions and hence the electrostatic environments experienced by each atom. Similar perturbations occur in liquids, although on a larger scale due to the translational, rotational and vibrational motions of the molecules. In gases the atoms or molecules are perturbed by collisions with other atoms or molecules, with the walls of the container,and with electrons if an ionising current is used. Whilst the last two of these may be controlled or eliminated, the rate of collisions at a given temperature between atoms or molecules of the gas itself is subject only to the pressure; the line-broadening it gives rise to is thus called either *collision-* or *pressure-broadening*.

Inhomogeneous line-broadening is principally caused in the solid state by the presence of impurity atoms in various crystallographic sites; a similar effect occurs in the case of liquid solutions. There is, however, a quite different inhomogeneous mechanism in gases. This results from the fact that the frequencies of the photons emitted are shifted by the well-known Doppler effect. If a molecule travelling with velocity v emits a photon of frequency ν in a direction k, as shown in Fig. 1.10, then the apparent frequency is given by

$$\nu' = \nu(1 + v_k/c')^{\frac{1}{2}}(1 - v_k/c')^{-\frac{1}{2}}$$
$$\approx \nu(1 + v_k/c'), \tag{1.24}$$

where v_k is the component of velocity in direction k, which is generally small compared to the speed of light c'. Since there is a Maxwellian distribution of molecular velocities, there is a corresponding range of Doppler-shifted frequencies given by Eq. (1.24). This effect is known as *Doppler broadening*.

As an illustration of the magnitudes of the various linewidths, a gas laser operating on an electronic transition at a pressure of 1 torr would have a natural linewidth typically of the order of 10 MHz, a collisional linewidth 100 MHz – 1 GHz, and a Doppler linewidth in the gigahertz range. For a similar laser operating on a vibrational transition, the natural linewidth would be a

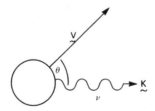

Fig. 1.10 Emission by a translating molecule. The velocity component v_k in the direction of emission is $v\cos\theta$

factor of about 10^5 smaller; hence, the natural linewidth is seldom the dominant line-broadening mechanism. Other line-broadening mechanisms are discussed at the beginning of Chap. 4, in connection with laser spectroscopy.

1.6
Properties of Laser Light and Their Applications

It has been recognized since the earliest days of laser technology that laser light has characteristic properties which distinguish it from that produced by other sources. In this section we shall look at how these properties arise from the nature of the lasing process and briefly consider examples of how they are utilized for particular laser applications. Such a discussion obviously has to be highly selective, and those examples which are presented here have been chosen more to illustrate the diversity of laser applications than for any other reason. Most specifically chemical uses are considered in detail in later chapters.

1.6.1
Beamwidth

Stimulated emission produces photons with almost precisely the same direction of propagation, the end-mirror configuration resulting in selective amplification of an axial beam that is typically only about 1 mm across. As there is no physical boundary to the beam, its radius or *waist* w is usually defined as the radial distance from the beam centre to points at which the intensity drops by a factor of e^2, i.e. to 13.5% of the central value. A typical figure for the angle of beam divergence is 1 mrad, sufficiently well directed to illuminate an area only 1 m across at a distance of 1 km; excimer lasers are available with beam divergences of less than 200 μrad. Although the extent of beam divergence is initially determined by the diffraction limit of the output aperture, the little divergence that there is can to a large extent be corrected by suitable optics.

A striking illustration of how well collimated a laser beam can be is provided by the fact that it has been possible to observe from Earth the reflection of laser light from reflectors placed on the surface of the moon by astronauts during the Apollo space programme. Other more prosaic applications hinging on laser collimation are *optical alignment* in the construction industry, as for example in tunnel boring, and *tracking and ranging* with pulsed lasers. In the latter case there are significant uses in connection with *atmospheric pollution monitoring*, about which we shall have more to say in Sect. 3.6.5. Here it is the narrow beamwidth that makes it possible to monitor from ground level, by analysis of scattered light, gases escaping from high factory chimneys.

The exact distribution of intensity within a laser beam is determined by the mode structure (see Sect. 1.3). In the simplest TEM_{00} case it consists of an essentially Gaussian distribution, as illustrated by the beam profile in Fig. 1.11 a. Ideally, any such beam of wavelength λ can be focussed to a spot

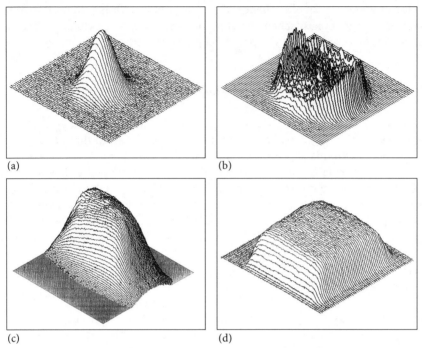

(a)   (b)

(c)   (d)

Fig. 1.11a–d Beam profiles (variation in intensity across the beam) for; **(a)** a He-Ne laser, **(b)** a diode laser, **(c)** an excimer laser, and **(d)** the same excimer beam as in (c), optically sculpted. (Reproduced by kind permission of Exitech Ltd.)

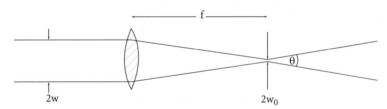

Fig. 1.12 Focussing a laser beam with a convex lens of focal length f

whose diffraction-limited waist $w_0 = 2\lambda/\pi\theta$, where θ is the angle of convergence as shown in Fig. 1.12. Attainment of the smallest focal spot clearly requires sharp focussing, since for a convex lens of focal length f we have $\theta = 2w/f$. In the common case of a laser beam which departs to some extent from the ideal Gaussian distribution – see Fig. 1.11. b–d – the formula for the focussed beam waist is often modified to read:

$$w_0 = 2M^2\lambda/\pi\theta, \tag{1.25}$$

where the parameter M^2 provides a good all-round measure of *beam quality*. For a fundamental Gaussian-mode beam, M^2 has a value of unity; in practice, values below 1.2 are regarded as optimum. Tight focussing is of particular importance in *microsampling* and *microprocessing* applications; it is also plays a key role in the production of high intensities (see following section).

1.6.2
Intensity

The property which is most commonly associated with laser light is a high intensity, and indeed lasers do produce the highest intensities known on Earth. Since a laser emits an essentially parallel beam of light in a well-defined direction, rather than in all directions, the most appropriate measure of intensity is the irradiance, as defined in Sect. 1.4.1. Since energy per unit time equals power, we thus have:

$$\text{Irradiance I} = \text{Power/Area.} \tag{1.26}$$

In using this equation, however, it must be stressed that 'power' refers to the output power, and not the input power of the laser. To put things into perspective when we look at typical laser irradiances, we can note that the mean intensity of sunlight on the Earth's surface is of the order of one kilowatt per square metre, i.e. 10^3 W m^{-2}.

Let us consider first a moderately powerful argon laser which can emit something like 10 W power at a wavelength of 488 nm. Assuming a cross-sectional area for the beam of 1 mm^2, this produces an irradiance of (10 W)/$(10^{-3}$ m$)^2 = 10^7$ W m^{-2}. In fact we can increase this irradiance by focussing the beam until we approach a diffraction limit imposed by the focussing optics. In this respect, too, laser light displays characteristically unusual properties, in that by focussing it is possible to produce intensities that exceed that of the source itself; this is not generally possible with conventional light sources. As a rough guide, the minimum radius of the focussed beam is comparable with the wavelength, so that in our example a cross-sectional area of 10^{-12} m^2 would be realistic and give rise to a focussed intensity of 10^{13} W m^{-2}.

Not surprisingly, however, it is in lasers which first accumulate energy as a population inversion is built up, and then release it through emission of a pulse of light, that we find the highest output intensities, though we have to remember that the peak intensity is obtained only for a very short time. A good Q-switched ruby laser, for example, which emits 25 ns pulses (1 ns = 10^{-9} s) at a wavelength of 694 nm, can give a peak output of 1 GW = 10^9 W in each pulse, though typically in a somewhat broad beam of about 500 mm^2 cross-sectional area. The mean irradiance of each pulse is thus approximately 2×10^{12} W m^{-2}, which can easily be increased by at least a factor of 10^6 by appropriate focussing. It should be noted that in all these rough calculations, it has been implicitly assumed that the intensity remains constant throughout the duration of each pulse, whereas in fact there is a definite rise at the beginning and decay at the end; in other words, there is a smooth *temporal profile*. Because the peak intensity from a pulsed laser is inversely proportional to its pulse duration, there are various methods of reducing pulse length so as to increase the intensity, and we shall examine these in Sect. 3.3.

Let us briefly look at a few applications of lasers which hinge on the high intensities available. A fairly obvious example from industry is *laser cutting and welding*. For such purposes, the high-power carbon dioxide and Nd-YAG lasers, which produce infra-red radiation, are particularly appropriate. Such lasers can cut through almost any material, though it is sometimes necessary to supply a jet of inert gas to prevent charring, for example with wood or paper; on the other hand, an oxygen jet facilitates cutting through steel. A focussed laser in the 10^{10} W m^{-2} range can cut through 3 mm steel at approximately 1 cm s^{-1}, or 3 mm leather at 10 cm s^{-1}, for example. Applications of this kind can be found across a wide range of industries, from aerospace to textiles, and there are several thousand laser systems being used for this purpose in the USA alone.

One of the most promising areas of medical applications is in *eye surgery*, for which several clinical procedures have already become quite well established. A detached retina, for example, which results in local blindness, can be 'spot welded' back onto its support (the choroid) by treatment with high intensity pulses of light from an argon laser. There are a great many advantages in the use of lasers for such surgery; the laser technique is non-invasive and does not require the administration of anaesthetics, nor the need for prolonged fixation by the eye during treatment, in view of the short duration of the pulses. When we return to a fuller discussion of pulsed laser systems in the next chapter, we shall see that there are far more exotic applications of high laser intensities than these.

1.6.3
Coherence

Coherence is the property which most clearly distinguishes laser light from other kinds of light, and it is, again, a property which results from the nature of the stimulated emission process. Light produced by more conventional thermal sources that operate by spontaneous emission is often referred to as being *chaotic*; there is generally no correlation between the phases of different photons, and appreciable intensity fluctuations result from the essentially random interference which ensues. By contrast, in the laser, photons emitted by the excited laser medium are emitted in phase with those already present in the cavity. The timescale over which phase correlation persists is known as the coherence time, t_c, and is given by

$$t_c = 1/\Delta\nu, \tag{1.27}$$

where $\Delta\nu$ is the emission linewidth. Directly related to this is the *coherence length*,

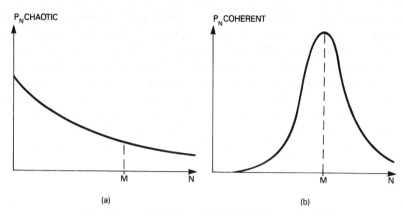

Fig. 1.13 a and b Photon statistics of (a) thermal radiation and (b) laser radiation of the same mean intensity

$$l_c = c't_c. \tag{1.28}$$

Thus two points along a laser beam separated by less than the coherence length should have related phase. The coherence length of a single mode output from a gas laser may be 100 m, but for a semiconductor laser it is more typically 1 mm. Measurement of the coherence length or coherence time of a laser is accomplished by *intensity fluctuation spectroscopy* and provides a useful means of ascertaining the emission linewidth (see Sect. 3.6.2).

Chaotic and coherent radiation have quite different *photon statistics*, as illustrated by the two graphs in Fig. 1.13. These graphs show the distribution of probability for finding N photons in a volume which contains on a time-average a mean number M. Chaotic light satisfies a Bose-Einstein distribution given by

$$P_N^{chaotic} = M^N/(M+1)^{N+1}, \tag{1.29}$$

whereas coherent light generally satisfies the Poisson distribution

$$P_N^{coherent} = M^N e^{-M}/N!. \tag{1.30}$$

Although most processes involving the interaction of light and matter cannot distinguish between the two kinds of light if they have the same mean irradiance (related to M by $M = IV/hc'\nu$, from Eq. 1.6), this is not the case with multiphoton processes, as we shall see in Sect. 4.6.

There are surprisingly few applications of laser coherence; the main one is *holography*, the technique for production of three-dimensional images. The process involves creating a special type of photographic image – a *hologram* – on a plate with a very fine photosensitive emulsion. Unlike the more usual kind of photographic image, the hologram contains information not only on the intensity, but also on the phase of light reflected from the subject; clearly, such an image cannot be created using a chaotic light source. Subsequent illumination of the image reconstructs a genuinely three-dimensional image. There are considerable difficulties to be overcome in creating true colour holograms, since phase information is lost if a range of wavelengths is used to create any single image. Although holograms which can be viewed in white light have become quite common, the colours they display are normally only the result of interference, not the colours of the original subject. Subject motion also tends to scramble phase information, so that the holographic creation of three-dimensional colour movies remains a tantalisingly distant prospect.

1.6.4
Monochromaticity

The last of the major characteristics of laser light, and the one which has the most relevance for chemical applications, is its essential monochromaticity, resulting from the fact that all of the photons are emitted as the result of a transition between the same two atomic or molecular energy levels, and hence have almost exactly the same frequency. Nonetheless, as we saw in Sect. 1.5.3, there is always a small spread to the frequency distribution, which may cover several discrete frequencies or wavelengths satisfying the standing-wave condition. The frequency interval between these allowed modes is called the *free spectral range*. As illustrated in Fig. 1.14, a number of closely separated frequencies may thus be involved in laser action, so that an additional means of frequency selection needs to be built into the laser to achieve the optimum degree of monochromaticity. What is generally used is an *etalon*, which is an optical element placed within the laser cavity and arranged so that only one

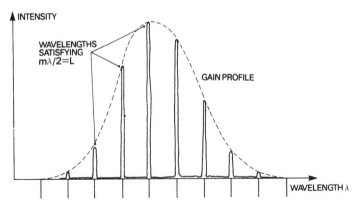

Fig. 1.14 Emission spectrum of a typical laser. Emission occurs only at wavelengths within the fluorescence linewidth of the laser medium which experience gain and also satisfy the standing-wave condition

well-defined wavelength can travel back and forth indefinitely between the end-mirrors.

With lasers which have a continuous output, it is quite easy to obtain an emission linewidth as low as 1 cm^{-1}, and in frequency-stabilized lasers the linewidth may be four or five orders of magnitude smaller. One of the important factors which characterize lasers is the quality factor Q, which equals the ratio of the emission frequency ν to the linewidth $\Delta\nu$,

$$Q = \nu/\Delta\nu. \tag{1.31}$$

The value of the Q-factor can thus easily be as high as 10^8, clearly of great significance for *high resolution spectroscopy*, which we shall examine at some length in Chap. 4. The spectroscopist often prefers to have linewidth expressed in terms of wavelength or wavenumber units, where the latter represents the number of wavelengths of radiation per unit length, usually per centimetre ($\tilde{\nu} = 1/\lambda$). Useful relationships between the *magnitudes* of the corresponding linewidth parameters are as follows:

$$\Delta\lambda = \lambda/Q; \tag{1.32}$$

$$\Delta\tilde{\nu} = \Delta\lambda/\lambda^2. \tag{1.33}$$

Another important area of technical application associated with the high degree of monochromaticity of a laser source lies in isotope separation. Since molecules which differ in isotopic constitution generally have slightly different absorption frequencies, by using a very narrow linewidth laser a mixture of the compounds can be selectively excited and then separated by other means. Not surprisingly, there is a great deal of interest in this kind of application in the nuclear industry. Again, a full discussion is reserved for later (Sect. 5.4).

1.7
Questions

1. Lasers usually emit a range of closely-spaced wavelengths satisfying the standing-wave condition of Eq. (1.1). Show that the smallest possible interval between wavelengths of light which will create standing waves inside a laser of cavity length L is $\Delta\lambda \approx \lambda^2/2L$, and derive an expression for the corresponding difference in wavenumber. Calculate the value of this interval for an argon ion laser having a cavity length of 1.5 m and operating at 488.0 nm.

2. Given that the mean number of round-trips of a photon inside a laser cavity is given by $(1 - R_1R_2)^{-1}$, where R_1 and R_2 are the reflectivities of the end-mirrors, calculate the mean time a photon spends inside a laser of cavity length 1 m and where $R_1R_2 = 90\%$ ($c = 3.00 \times 10^8$ m s^{-1}).

3. Estimate the mean photon retention time for the world-class Canterbury (NZ) ring dye laser, in which photons circulate in a one-way trip around a

cavity of length 3.477 m, successively reflected by four mirrors whose product reflectivity exceeds 99.999% ($c = 3.00 \times 10^8$ m s^{-1}).

4. A helium-neon laser beam (632.8 nm) with a near-Gaussian intensity distribution, characterised by a beam waist of 0.4 mm and a beam quality factor $M^2 = 1.1$, is passed through a convex lens of focal length 18 mm. Calculate the angle of convergence and the minimum beam waist at focus.

5. Calculate the quality factors Q for (a) a semiconductor diode laser emitting at a wavelength of 1 μm and with an emission linewidth of 10^{-3} cm^{-1}, and (b) a TEA carbon dioxide laser emitting at 1000 cm^{-1} with a linewidth of 10 cm^{-1}.

6. A sample of carbon disulphide (molar volume 6.04×10^{-5} m^3 mol^{-1}) is irradiated by a Q-switched ruby laser (operational wavelength 694 nm) giving peak intensities of 1.6×10^{16} W m^{-2}. At the moment of peak intensity, what is the average number of photons travelling through the region of space occupied by one molecule?
($h = 6.63 \times 10^{-34}$ J s : $c = 3.00 \times 10^8$ m s^{-1} : $L = 6.02 \times 10^{23}$ mol^{-1}).

7. With any kind of electromagnetic radiation there is associated an oscillatory electric field E which induces a fluctuating electric dipole moment even in centrosymmetric species such as atoms. The magnitude of this dipole is given by $\mu_{induced} = 3\varepsilon_0 E V_m L^{-1}(n^2 - 1)/(n^2 + 2)$, where V_m is the molar volume and n the refractive index at the appropriate wavelength. Calculate the dipole moment induced in an argon atom at the focus of a giant pulse laser where the magnitude of E reaches a value of 10 V nm^{-1}. If this dipole were considered to be due to two charges $+e$ and $-e$, what would be their separation? ($n(Ar) = 1.000281$: $V_m(Ar) = 2.24 \times 10^{-2}$ m^3 mol^{-1} : $L = 6.02 \times 10^{23}$ mol^{-1} : $\varepsilon_0 = 8.85 \times 10^{-12}$ F m^{-1} : $e = 1.60 \times 10^{-19}$ C : $1\,D = 3.34 \times 10^{-30}$ C m; note $1\,V \times 1\,F = 1\,C$).

8. The Maxwell-Bartoli equation, derived in the nineteenth century, provides a means of calculating the light-pressure p on an object of reflectivity R in a beam of light of irradiance I: the equation reads $p = (R + 1)I/c$. Using this result, estimate the mass of a perfectly reflective particle which can be levitated in vacuo by a 1 kW cw CO_2 laser beam. ($c = 3.00 \times 10^8$ m s^{-1}; $g = 9.81$ m s^{-2}).

CHAPTER 2

Laser Sources

To ... add another hue unto the rainbow ... is wasteful
and ridiculous excess
'King John', William Shakespeare

Since the construction of the first laser based on ruby, widely ranging materials have been adopted as laser media, and the range is still continually being extended to provide output at new wavelengths: according to Charles H. Townes, one of the pioneers of laser development, 'almost anything works if you hit it hard enough'. Naturally occurring laser emission has even been observed in the clouds of hydrogen gas surrounding a distant star. As can be seen from a glance at Appendix 1, output from commercial lasers now covers most of the electromagnetic spectrum through from the microwave region to the ultraviolet, and much effort is being concentrated on extending this range to still shorter wavelengths. Amongst a host of tantalising possibilities is the prospect of holographically imaging molecules by use of an X-ray laser, for example.

Having discussed the general principles and characteristics of laser sources in the last chapter, we can now have a look at some of the specific types of laser commercially available and commonly used, both within and outside of chemistry. Space does not permit a thorough description of the large number of lasers that fall into this category, and the more limited scope of this chapter is therefore a discussion of representative examples from each of the major types of laser.

2.1
Optically Pumped Solid-State Lasers

With the exception of semiconductor lasers, which we shall deal with separately, the active medium in solid-state lasers is generally a transparent crystal or glass into which a small amount of transition metal is doped; transitions in the transition metal ions are responsible for the laser action. The two most common dopant metals are chromium, in the ruby and alexandrite lasers, and neodymium, in the Nd:YAG and Nd:glass lasers, and the dopant concentration is typically 1% or less. All such lasers are optically pumped by a broadband flashlamp source and can be pulsed to produce the highest available laser intensities. A natural irregularity in the temporal profile of each pulse known as *spiking* makes this kind of laser particularly effective for cutting and drilling applications.

2.1.1
Ruby Laser

The most familiar solid-state laser is the ruby laser, which has an important place in the history of lasers as it was the first type ever constructed, in 1960. In a beautiful example of the ironies of science, the inventor Theodore H. Maiman had his original paper describing this first laser turned down for publication by *Physical Review Letters*, because it was deemed to be of insufficient interest. In construction, the modern ruby laser comprises a rod of commercial ruby (0.05% Cr_2O_3 in an Al_2O_3 lattice) of between 3 and 25 mm in diameter and up to 20 cm in length. The chromium ions are excited by the broadband emission from a flashlamp coiled around it, or placed alongside it within an elliptical reflector as shown in Fig. 2.1.

Before we consider the details of the energy level scheme for the ruby laser, it is interesting to note that we can learn something straightaway from the colour of light it emits. The fact that a ruby laser emits red light might not seem too surprising, until we remember that the reason ruby appears red is because it *absorbs* in the green and violet regions of the spectrum; that is why it *transmits* (or reflects) red light. So since the absorption from the flashlamp and the laser emission evidently occur at different wavelengths, it is immediately clear that ruby is a laser involving more than two levels, as we should expect from our discussion in Chap. 1. In fact, the energy level diagram is as

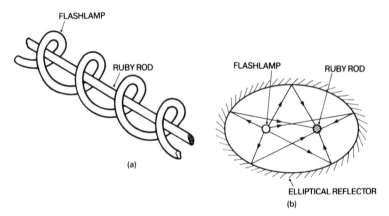

Fig. 2.1a and b Arrangement of flashlamp tube and ruby rod in a ruby laser; (a) illustrates the helical flashlamp option, and (b) the highly efficient pumping obtained from a confocal linear flashlamp and laser rod within an elliptical reflector

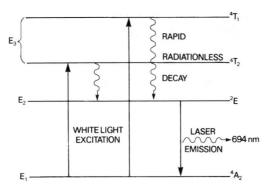

Fig. 2.2 Energetics of the ruby laser

shown in Fig. 2.2, and may be regarded as a *pseudo-three-level system*, in the sense that in the course of excitation, decay and emission, only three Cr^{3+} levels are involved; $^4A_2$, $^4T_1$ or $^4T_2$, and ^2E (cf. Fig. 1.7). It is worth noting that these energy levels are quite different from those of a free chromium atom. The electrostatic environment created by the surrounding atoms of the host lattice, known as the *crystal field*, produces large splittings between energy levels which in the free atom would normally be degenerate.

The initial flashlamp excitation takes the Cr^{3+} ions up from the ground state E_1 ($^4A_2$) to one of the two E_3 (^4T) levels. Both of these levels have sub-nanosecond lifetimes, and rapidly decay into the E_2 level (^2E). This decay takes place simply by non-radiative processes which channel energy into lattice vibrations and so result in a heating up of the crystal. Since the E_2 level has a much longer decay time of around 4 ms, a population inversion is created between the E_2 and E_1 states, leading to laser emission at a wavelength of 694.3 nm as the majority of Cr^{3+} ions simultaneously cascade down to the ground state. The lasing process thus generates a single, intense pulse of light, typically of between 0.3 and 3 ms in duration, and it is necessary to recharge the flashlamp before the next pulse can be created. The delay between successive pulses usually lasts several seconds and can be as long as a full minute. A very different kind of pulsing can be created by the technique known as *Q-switching*, which is discussed in Sect. 3.3.2.

One problem with the ruby laser, common to all lasers of this type, is the damage caused by the repeated cycle of heating and cooling associated with the generation of each pulse, which ultimately necessitates replacement of the ruby rod. To improve performance, the rod usually needs to be cooled by circulation of water in a jacket around it. Despite its drawbacks, with pulse energies as high as 200 J the ruby laser represents a very powerful source of monochromatic light in the optical region, and it has found many applications in materials processing. The emission bandwidth is typically about 0.5 nm (10 cm), but this can be reduced by a factor of up to 10^4 by introducing intracavity etalons. With such narrow linewidth, the laser is then suitable for holography, and this is where most of its applications are now found. Another field of application is in lidar (Sect. 3.6.5). The beam diameter of a low-power ruby laser can be as little as 1 mm, with 0.25-mrad divergence; the most powerful lasers may have beams up to 25 mm in diameter, and a larger divergence of several milliradians.

2.1.2
Neodymium Lasers

Neodymium lasers are of two main types; in one, the host lattice for the neodymium ions is a crystal, usually yttrium aluminium garnet crystal ($Y_3Al_5O_{12}$), and in the other the host is an amorphous glass. These are referred to as Nd:YAG and Nd:glass, respectively. Although transitions in the neodymium ions are responsible for laser action in both cases, the emission characteristics differ because of the influence of the host lattice on the neodymium energy levels. Also, glass does not have the excellent thermal conductivity properties of YAG crystal, so that it is more suitable for pulsed than for continuous-wave (cw) operation. Like the ruby laser, both types of neodymium laser are normally pumped by a flashlamp arranged confocally alongside a rod of the laser material in an elliptical reflector. The very small hand-held YAG lasers now available are pumped by diode lasers (see the end of this section and Sect. 2.2).

Once again, the energy levels of neodymium ions involved in laser action, which are naturally degenerate in the free state, are split by interaction with the crystal field. In this case, the splitting is readily illustrated in schematic form, as in Fig. 2.3. As a result, transitions between components of the $^4F_{3/2}$ and $^4I_{11/2}$ states, which are forbidden in the free state, become allowed and can give rise to laser emission. The $^4F_{3/2}$ levels are initially populated following non-radiative decay from higher energy levels excited by the flashlamp, and since the terminal $^4I_{11/2}$ laser level lies above the $^4I_{9/2}$ ground state, we thus have a *pseudo-four-level system* (cf. Fig. 1.8).

The principal emission wavelength for both types of neodymium laser is around 1.064 μm, in the near-infra-red (or 1.053 μm in the less common Nd:YLF lasers, where lithium replaces aluminium in the host lattice). Some commercial lasers can also be operated on a different transition producing 1.319 μm output. The YAG and glass host materials impose very different characteristics on the emission, however. Quite apart from the differences in thermal conductivity which determine the question of continuous or pulsed operation, one of the main differences appears in the linewidth. Since glass has an amorphous structure, the electrical environment experienced by different neodymium ions varies, and so the crystal field splitting also varies from ion to ion. Because of this, the linewidth tends to be much broader than in the Nd:YAG laser, where the lattice is much more regular and the field splitting more constant. However, the concentration of neodymium dopant in glass

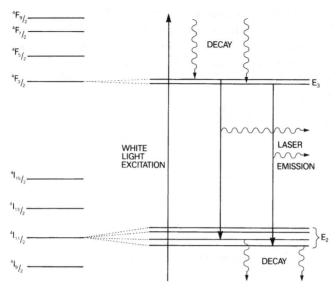

Fig. 2.3 Energetics of a neodymium laser

may be as large as 6%, compared with 1.5% in a YAG host, so that a much higher energy output can be obtained. For both these reasons, the Nd:glass laser is ideally suited for the production of extremely high intensity ultrashort pulses by the technique of mode-locking (see Sect. 3.3.3). In fact it is with the neodymium lasers that we find the greatest available laser beam intensities.

The output power of a typical continuous-wave Nd:YAG laser is several watts and can exceed 200 W. When operating in a pulsed mode, the energy per pulse varies according to the method of pulsing and the pulse repetition rate, but can be anywhere between a small fraction of a joule and 100 J for a single pulse. Such powerful sources of infra-red radiation now find many applications in materials processing. Nd:YAG lasers have also made substantial inroads into the surgical laser market. Whilst use is often made of localised thermally induced processes caused through the absorption of intense infra-red radiation, there is much interest in the alternative *breakdown-mode surgery* possible with Q-switched or mode-locked YAG lasers. Here, the enormous electric field (typically 3×10^8 V m^{-1}) associated with each focussed pulse of laser light strips electrons from tissue molecules, and the resultant plasma creates a shock wave which causes mechanical rupture of tissue within a distance of about 1 mm from the focus. This method is proving highly useful for a number of ophthalmic microsurgical applications. It is important to note that the majority of Nd:YAG research applications in photochemistry and photobiology make use not of the 1.064 μm radiation as such, but rather the high intensity visible light which can be produced by frequency conversion methods. Particularly important in this respect are the wavelengths of 532, 355 and 266 nm obtained by harmonic generation (see Sect. 3.2.2).

Diode-pumped solid-state lasers represent a comparatively new and very different laser technology. These miniature-scale devices, which can often fit comfortably into the hand, incorporate a semiconductor diode laser (see Sect. 2.2) to directly pump a small Nd:YAG crystal. Modifications to the cavity configuration enable linewidths of less than 5 kHz (2×10^{-7} cm^{-1}) to be attained in certain models. Whilst cw output powers are generally low (on the milliwatt scale), the crucial 1W target can be exceeded in some commercially available systems, and Q-switched pulsing should offer the possibility of multi-kilowatt peak output. A perhaps more important consideration is the common inclusion of frequency-conversion crystals in these devices, so providing milli-watt emission in the visible region, especially at the 532 nm line in the green. Such integrated solid-state lasers offer the advantages of smaller size, higher efficiency, better stability and lower noise levels than most gas discharge lasers, and they also offer the prospect of a more standard format for ancillary equip-ment than has been possible in the past. Applications cover a wide spectrum of optical instrumentation, and materials processing capabilities have also been demonstrated.

2.1.3
Tunable Lasers

Recently, other types of solid-state laser have been developed which, although again based on optical transitions in transition metal ions embedded in a host ionic crystal, produce tunable radiation. For a number of spectroscopic appli-cations, such lasers represent an increasingly attractive alternative to the more traditional tunable lasers based on organic dyes (Sect. 2.5). Their chief advan-tages are an appreciably more rugged and compact construction and operation without use of toxic chemicals.

A good example is provided by the *alexandrite* laser, in which the active medium is 0.01%–0.4% chromium-doped alexandrite (BeAl$_2$O$_4$) crystal. Cru-cial differences arise between ruby and alexandrite lasers as a result of the fact that the Cr^{3+} electronic ^{4}A$_2$ ground state is in the latter case no longer discrete, but because of coupling with lattice vibrations consists of a broad continuous band of *vibronic* energy levels. Laser emission, following the usual flashlamp excitation, can in this case take place through downward transitions from the ^{4}T$_2$ state to anywhere in the ground state continuum, and the result is tunable output covering the region 700–825 nm (though not across the entire range without changing operating conditions). For this reason, the alexandrite laser is often referred to as a *vibronic laser*.

The recently developed titanium:sapphire laser, commonly pumped by a primary laser beam such as the all-lines output of an ion gas laser, falls into a similar category. Based on Al$_2$O$_3$ crystal doped with approximately 0.1% Ti$_2$O$_3$, lasing here occurs on vibronic $^2E \rightarrow {}^2T_2$ transitions in the Ti^{3+} ions, offering high-power tunable cw or pulsed emission over the exceptionally large wavelength range 650–1100 nm in the near-infra-red – though the cavity end-mirrors need changing to cover the whole range. At around 800 nm, over 5 W cw outputs can be achieved. The titanium:sapphire laser is finding an increasingly wide range of applications, particularly in spectroscopy. Other commercially available vibronic lasers offering pulsed output in the infra-red are Co^{2+}-doped MgF_2 (the first tunable solid-state laser) tunable over the range 1.75–2.50 μm, and chromium-doped Mg_2SiO_4 (forsterite) in which, although Cr^{3+} is present the laser-active species is Cr^{4+}, and the corresponding emission range is 1.15–1.35 μm.

One other type of solid-state laser in which the active medium is a regular ionic crystal may be mentioned here, although it does not involve transition metal ions. This is the so-called *F-centre* (or *colour centre*) laser, which operates on optical transitions at defect sites in alkali halide crystals, as, for example, in KCl doped with thallium; the colour centres are typically excited by a pump Nd:YAG or argon/krypton ion laser. Such lasers produce radiation which is tunable using a grating end mirror over a small range of wavelengths in the overall region 0.8–3.4 μm; different crystals are required for operation over different parts of this range. One disadvantage of this type of laser is that it requires use of liquid nitrogen since the crystals need to be held at cryogenic temperatures. Colour centre lasers have found limited applications, although they have proven use in the high-resolution spectroscopy of small molecules and molecular fragments.

2.2
Semiconductor Lasers

In the solid-state lasers we have considered so far, the energy levels are associated with dopant atoms in quite low concentrations in a host lattice of a different material. Because under these conditions the dopant atoms are essentially isolated from each other, their energy levels remain discrete, and we obtain the same kind of line spectrum which we generally associate with isolated atoms or molecules. In the case of semiconductor lasers, however, we are dealing with energy levels of an entire lattice, and so it is energy bands rather than discrete energy levels which we have to consider.

The characteristic properties of a semiconductor arise from the fact that there is a small energy gap between two energy bands known as the valence band and the conduction band. At very low temperatures, the electrons associated with the valence shell of each atom occupy energy levels within the valence band, and the higher energy conduction band is empty. Hence electrons are not able to travel freely through the lattice, and the material has the electrical properties of an insulator. At room temperature, however, thermal energy is sufficient for some electrons to be excited into the conduction band where passage through the lattice is relatively unhindered, and hence the electrical conductivity rises to somewhere inbetween that expected of a conductor and an insulator.

Solid-state electronic devices generally make use of junctions between p-type and n-type semiconductor. The former type has in its lattice impurity atoms which possess fewer valence electrons than the atoms they replace; the latter has impurity atoms possessing more valence electrons than the atoms they replace. The most familiar semiconductor materials are Class IV elements such as silicon and germanium. However, binary compounds between Class III and Class V elements, such as gallium arsenide (GaAs), exhibit similar behaviour. In this case, the lattice consists of two interpenetrating face-centred cubic lattices, and p- and n-type crystal are obtained by varying the stoichiometry from precisely 1:1. In the p-type material, some of the arsenic atoms are replaced by gallium; in n-type crystal, the converse is true.

Semiconductor lasers operate in a broadly similar way to the more familiar light-emitting diodes (LED's) widely used in electronic gadgets. By applying an electrical potential across a simple diode junction between p- and n-type crystal, electrons crossing the semiconductor boundary drop down from a conduction band to a valence band, emitting radiation in the process as illustrated in Fig. 2.4. The emission is most often in the infra-red, and the optical properties of the crystal at such wavelengths make it possible for the crystal end-faces to form the confines of the resonant cavity. One advantage of this kind of laser is its extremely small size, which is usually about half a millimetre. However, this does result in very poor beam quality (see the typical beam profile shown in Fig. 1.11b, page 18) and poor collimation; divergences of 10° are by no means unusual, creating the need for corrective optics in many applications.

There are two main types of semiconductor laser: those operating at fixed wavelengths, and those which are tunable. The three most common fixed-wa-

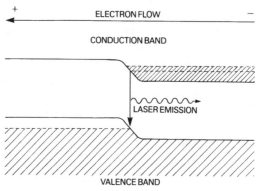

Fig. 2.4 Energetics of a diode laser; the energy levels of the p-type material are shown on the left, and the n-type on the right

velength types are gallium arsenide, gallium aluminium arsenide, and indium gallium arsenide phosphide. Gallium arsenide lasers emit at a wavelength of around 0.904 μm; the wide range of flexibility in the stoichiometry of the other types makes it possible to produce a range of lasers operating at various fixed wavelengths in the region 0.8–1.3 μm. The so-called 'lead salt' diode lasers, however, which are derived from non-stoichiometric binary compounds of lead, cadmium and tin with tellurium, selenium and sulphur, emit in the range 2.8–30 μm (3500–330 cm^{-1}), depending on the exact composition. Although these lasers require a very low operating temperature, typically in the range 15–90 K, the operating wavelength is very temperature-dependent, and so the wavelength can be tuned by varying the temperature. The tuning range for a lead salt diode laser of a particular composition is typically about 100 cm^{-1}.

Modes in a diode laser[1] are typically separated by 1–2 cm^{-1}, and each individual mode generally has a very narrow linewidth, of 10^{-3} cm^{-1} or less. The output power of continuous semiconductor lasers is generally measured in milliwatts, although some array devices are capable of emitting as much as 10 W. By far the most important applications lie in the area of optical pickups such as those in compact disc players, optical communications and fibre optics; not surprisingly, diode lasers currently represent the most rapidly growing area of the laser market. However, diode lasers are also very well suited to high-resolution infra-red spectroscopy, since the linewidth is sufficiently small to enable the rotational structure of vibrational transitions to be resolved for many small molecules (see Sect. 4.2). This method has proved particularly valuable in characterising short-lived intermediates in chemical reactions. More diverse applications can be expected as the range of diode laser emission wavelengths encroaches further into the visible region – devices emitting at the blue end of the spectrum are now available, in some cases through frequency doubling methods (see Sect. 3.2.2). As prices fall, diode lasers may be expected to oust helium-neon lasers from many of their traditional roles.

2.3
Atomic and Ionic Gas Lasers

The class of lasers in which the active medium is a gas covers a wide variety of devices. Generally, the gas is either monatomic, or else it is composed of very simple molecules. Representative examples of monatomic lasers are discussed in this section; molecular gas lasers are discussed in Sect. 2.4. In both cases, since laser emission results from optical transitions in *free* atoms or molecules, usually at low pressures, the emission linewidth can be very small. The gas is often contained in a sealed tube, with the initial excitation provided by an elec-

1) In quantum microcavity semiconductor lasers there can be just one mode supported by the cavity

trical discharge, so that in many cases the innermost part of the laser bears a superficial resemblance to a conventional fluorescent light.

The laser tube can be constructed from various materials and need not necessarily be transparent. Unfortunately metals are generally ruled out because they would short-circuit the device. Silica is commonly employed, and also beryllium oxide which has the advantage for high power sources of a high thermal conductivity. It is quite a common feature to have the laser tube contain a mixture of two gases, one of which is involved in the pump step and the other in the laser emission. Such gas lasers are usually very reliable, since there is no possibility of thermal damage to the active medium which there is for solid-state lasers, and for routine purposes they are the most widely used type.

2.3.1
Helium-Neon Laser

The helium-neon laser was the first cw laser ever constructed, and it was also the first laser to be made available to a commercial market, in 1962. The active medium is a mixture of the two gases contained in a glass tube at low pressure; the partial pressure of helium is approximately 1 mbar and that of neon 0.1 mbar. The initial excitation is provided by an electrical discharge and serves primarily to excite helium atoms by electron impact. The excited helium atoms subsequently undergo a process of collisional energy transfer to neon atoms; it so happens that certain levels of helium and neon are very close in energy, so that this transfer takes place with a high degree of efficiency. Because the levels of neon so populated lie above the lowest excited states, a population inversion is created relative to these levels, enabling laser emission to occur as shown in Fig. 2.5. Two points about this should be made in passing. First, note that the usual state designation cannot be given for the energy levels of neon, because Russell-Saunders coupling does not apply. Secondly, each electron configuration gives rise to several closely spaced states, but only those directly involved in the laser operation have been shown on the diagram.

Three distinct wavelengths can be produced in the laser emission stage; there is one visible wavelength, typically of milliwatt power, which appears in the red at 632.8 nm, and there are two infra-red wavelengths of somewhat lower power at 1.152 and 3.391 μm. Obviously, infra-red optics are required for operation of the laser at either of these wavelengths. Following emission, the lasing cycle is completed as the neon undergoes a two-step radiationless decay back down to its ground state. This involves transition to a metastable $2p^53s^1$ level, followed by collisional deactivation at the inner surface of the tube. The last step has to be rapid if the laser is to operate efficiently; for this reason, the surface/volume ratio of the laser tube has to be kept as large as possible, which generally means keeping the tube diameter small. In practice, tubes are commonly only a few millimetres in diameter. Other very weak transitions have been utilised in the production of a 1-mW helium-neon laser emitting at var-

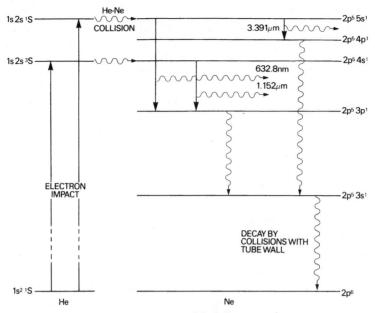

Fig. 2.5 Energetics of the helium-neon laser

ious wavelengths including 543.5 nm in the green; the principal virtue of this laser is that it is substantially cheaper than any other green laser.

Helium-neon lasers operate continuously, and despite their low output power, they have the twin virtues of being both small and relatively inexpensive. Consequently, they can be found in more widely ranging applications than any other kind of laser. Their principal applications are those which hinge on the typically narrow laser beamwidth, but where power is not too important. Examples include many optical scanning devices used for quality control and measurement in industry; helium-neon scanners are also used in optical video disc systems, supermarket bar-code readers and optical character recognition equipment. Other uses include electronic printing and optical alignment. One other laser of a similar type is the helium-cadmium laser, in which transitions in free cadmium atoms result in milliwatt emission at 442 nm in the blue and 325 nm in the ultraviolet. The blue line is particularly well suited to high-resolution applications in the printing and reprographics industry.

2.3.2
Argon Laser

The argon laser is the most common example of a family of ion lasers in which the active medium is a single-component inert gas. The gas, at a pressure of approximately 0.5 mbar, is contained in a plasma tube of 2- to 3-mm bore and

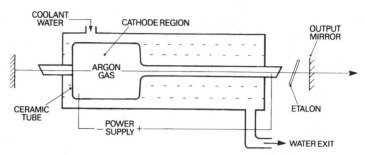

Fig. 2.6 Schematic construction of an argon ion laser. Magnetic coils (not shown) may be placed around the discharge tube to keep the plasma confined to the centre of the bore

is excited by a continuous electrical discharge (see Fig. 2.6). Argon atoms are ionised and further excited by electron impact; the nature of the pumping process produces a population of several ionic excited states, and those responsible for laser action are on average populated by two successive impacts. The sustainment of a population inversion between these states and others of lower energy results in emission at a series of discrete wavelengths over the range 350–530 nm (see Appendix 1), the two strongest lines appearing at 488.0 and 514.5 nm. These two wavelengths are emitted as the result of transitions from the singly ionised states with electron configuration $3s^2 3p^4 4p^1$ down to the $3s^2 3p^4 4s^1$ state. Further radiative decay to the multiplet associated with the ionic ground configuration $3s^2 3p^5$ then occurs, and the cycle is completed either by electron capture or further impact excitation. Doubly ionised Ar^{2+} ions contribute to the near-ultraviolet laser emission.

Since several wavelengths are produced by this laser, an etalon or dispersing prism is generally placed between the end-mirrors to select one particular wavelength of light for amplification; the output wavelengths can thus be varied by changing its orientation. By selecting a single longitudinal mode, an output linewidth of only 0.0001 cm^{-1} is obtainable. The pumping of ionic levels required for laser action requires a large and continuous input of energy and the relatively low efficiency of the device means that a large amount of thermal energy has to be dispersed. Cooling therefore has to be a major consideration in the design, and as shown in Fig. 2.6, circulation of water in a jacket around the tube is the most common solution, although air-cooled argon lasers are available. The output power of a cw argon laser usually lies in the region running from milliwatts up to about 25 W.

Argon lasers are fairly expensive, comparatively fragile, and have tube lifetimes generally limited to between 1000 and 10 000 hours. One of the principal reasons for the limited lifetime is erosion of the tube walls by the plasma, resulting in deposition of dust on the Brewster output windows. The argon itself is also depleted to a small extent by ions becoming embedded in the tube walls. Despite these drawbacks, such lasers have found widespread research applications in chemistry and physics, particularly in the realm of spectroscopy, where they are often employed to pump dye lasers (see Sect. 2.5). Argon lasers have also made a strong impact in the printing and reprographics industry, in entertainment and visual displays, and they play an increasingly important role in medical treatment, especially in laser ophthalmological procedures.

The other common member of the family of ion lasers is the krypton laser. In most respects it is very similar to the argon laser, emitting wavelengths over the range 350–800 nm, although because of its lower efficiency the output takes place at somewhat lower power levels, up to around 5 W. The strongest emission is at a wavelength of 647.1 nm. In fact, the strong similarity in physical requirements and performance between the argon and krypton lasers makes it possible to construct a laser containing a mixture of the two gases, which provides a very good range of wavelengths across the whole of the visible spectrum, as can be seen from Appendix 1. Such lasers emit many of the wavelengths useful for biomedical applications, the blue-green argon lines being especially useful since they are strongly absorbed by red blood cells.

2.3.3
Copper Vapour Laser

The copper vapour laser has not made a strong impact on the laser market, though it does have a number of features which should make it a very attractive competitor for a number of applications. It beongs to the class of *metal-*

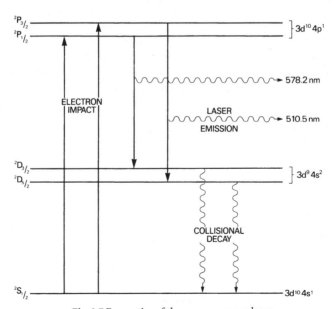

Fig. 2.7 Energetics of the copper vapour laser

vapour lasers, in which transitions in free uncharged metal atoms give rise to laser emission.

The copper laser is essentially a three-level system involving the energy levels shown in Fig. 2.7. Electron impact on the ground state copper atoms results in excitation to ^{2}P states belonging to the electron configuration $3d^{10}\,4p^1$, from which transitions to lower-lying $3d^9 4s^2\ {}^2$D levels can take place. Laser emission thus occurs at wavelengths of 510.5 nm in the green and 578.2 nm in the yellow. Further collisions of the excited atoms with electrons or the tube walls subsequently result in decay back to the ground state. One of the problems with this particular scheme is that electron impact on ground state copper atoms not only populates the ^{2}P levels but also the ^{2}D levels associated with the lower end of the laser transitions. For this reason, it is not possible to sustain a population inversion between the ^{2}P and ^{2}D levels and so the laser naturally operates in a pulsed mode, usually with a pulse repetition frequency of about 5 kHz. Each pulse typically has a duration of 30 ns and an energy in the millijoule range.

The physical design of the laser involves an alumina plasma tube containing small beads or other sources of metallic copper at each end. The tube also contains a low pressure of neon gas (approximately 5 mbar) to sustain an electrical discharge. Passage of a current through the tube creates temperatures of 1400–1500°C, which heats the copper and produces a partial pressure of Cu atoms of around 0.1 mbar; this then acts as the lasing medium. One of the initial drawbacks of the laser, the long warm-up time of about an hour, has now been overcome in a variation on this design which operates at room temperature.

The chief advantages of the copper vapour laser are that it emits visible radiation at very high powers (the time average over a complete cycle of pulsed emission and pumping is 10–60 W), has a high repetition rate, and that it is reasonably priced and highly energy-efficient. The most powerful 100-W copper vapour laser consumes only half as much input power as a 20-W argon

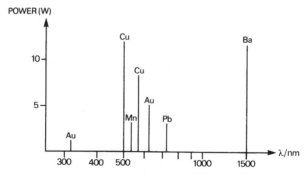

Fig. 2.8 Typical metal vapour output powers. (Reproduced by kind permission of Quentron Optics Pty. Ltd.)

laser, for example. Applications of the copper laser, which are at present mostly at the research and development stage, include uranium isotope separation (see Sect. 5.4.1). There has also been interest in use of the laser for photography and holography, as well as in underwater illumination, for which the emission wavelengths are particularly well suited to minimise attenuation. Dermatological applications are also being investigated, since the 578-nm emission lies helpfully close to the haemoglobin absorption peak at 577 nm.

Other metal-vapour lasers that have graduated from the research to the production stage are the barium laser, emitting at 1.50, 2.55 and 4.72 μm, and the gold laser, emitting a few watts power principally at a wavelength of 628 nm. The latter has already proved highly effective for cancer phototherapy (see Sect. 5.6). Typical output powers and wavelengths available from metal vapour lasers are shown in Fig. 2.8.

2.4
Molecular Gas Lasers

2.4.1
Carbon Dioxide Laser

The carbon dioxide laser is our first example of a laser in which the transitions responsible for stimulated emission take place in free molecules. In fact, the CO_2 energy levels involved in laser action are not electronic but rotation-vibration levels, and emission therefore occurs at much longer wavelengths, well into the infra-red. The lasing medium consists of a mixture of CO_2, N_2 and He gas in varying proportions, but often in the ratio 1:4:5; the helium is added to improve the lasing efficiency, and the nitrogen plays a role similar to that of helium in the He-Ne laser.

The sequence of excitation is illustrated in Fig. 2.9. The first step is population of the first vibrationally excited level of nitrogen by electron impact. In its ground vibrational state each nitrogen molecule can possess various discrete amounts of rotational energy, and various rotational sub-levels belonging to the vibrationally excited state are populated by the electron collision. These levels are all metastable, since radiative decay back down to the vibrational ground state is forbidden by the normal selection rules for emission. However, one of the vibrationally excited states of carbon dioxide, labelled (001) since it possesses one quantum of energy in the ν_3 vibrational mode (the antisymmetric stretch) has almost exactly the same energy as the vibrationally excited nitrogen molecule. Consequently, collision between the two molecules results in a very efficient transfer of energy to the carbon dioxide; once again, it is in fact the rotational sub-levels belonging to the (001) state which are populated.

Laser emission in the CO_2 then occurs by two routes, involving radiative decay to rotational sub-levels belonging to the (100) and (020) states; the former possess one quantum of vibrational energy in the ν_1 symmetric stretch

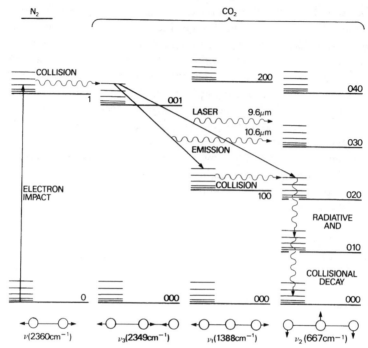

Fig. 2.9 Energetics of the carbon dioxide laser. The rotational structure of the vibrational levels is shown only schematically

mode, while the latter have two quanta of energy in the ν_2 bending mode. These states cannot be populated directly by collision with N_2, so that they exist in a population inversion with respect to the (001) levels. The two laser transitions result in emission at wavelengths of around 10.6 μm and 9.6 μm, respectively. Both pathways of decay ultimately lead to the (020) states, as shown in the diagram; subsequent deactivation results both from radiative decay and from collisions with He atoms. Since various rotational sub-levels can be involved in the emissive transitions, use of suitable etalons enable the laser to be operated at various discrete frequencies within the 10.6 μm and 9.6 μm bands as listed in Table 2.1; in wavenumber terms a typical gap between successive emission frequencies is 2 cm^{-1}. Since the positions of the CO_2 energy levels vary for different isotopic species, other wavelengths may be obtained by using isotopically substituted gas.

One of the problems which must be overcome in a carbon dioxide laser is the possibility that some molecules will dissociate into carbon monoxide and oxygen in the course of the excitation process. When a sealed cavity is used, this problem is generally overcome by the admixture of a small amount of water vapour, which reacts with any carbon monoxide and regenerates carbon dioxide. Such a method is unnecessary if cooled carbon dioxide is continu-

Table 2.1 Principal emission lines of the carbon dioxide laser
The emission lines, whose frequencies are expressed in wavenumber units $\tilde{\nu}$ below, result from rotation-vibration transitions in which the rotational quantum number J changes by one unit. Transitions from J $\rightarrow$ J – 1 are denoted by P(J); transitions from J $\rightarrow$ J + 1 are denoted by R(J).

10.6 µm band (001 $\rightarrow$ 100 transitions)

Line	$\tilde{\nu}$ (cm^{-1})	Line	$\tilde{\nu}$ (cm^{-1})	Line	$\tilde{\nu}$ (cm^{-1})
P(56)	907.78	P(22)	942.38	R(18)	974.62
P(54)	910.02	P(20)	944.19	R(20)	975.93
P(52)	912.23	P(18)	945.98	R(22)	977.21
P(50)	914.42	P(16)	947.74	R(24)	978.47
P(48)	916.58	P(14)	949.48	R(26)	979.71
P(46)	918.72	P(12)	951.19	R(28)	980.91
P(44)	920.83	P(10)	952.88	R(30)	982.10
P(42)	922.92	P(8)	954.55	R(32)	983.25
P(40)	924.97	P(6)	956.19	R(34)	984.38
P(38)	927.01	P(4)	957.80	R(36)	985.49
P(36)	929.02	R(4)	964.77	R(38)	986.57
P(34)	931.00	R(6)	966.25	R(40)	987.62
P(32)	932.96	R(8)	967.71	R(42)	988.65
P(30)	934.90	R(10)	969.14	R(44)	989.65
P(28)	936.80	R(12)	970.55	R(46)	990.62
P(26)	938.69	R(14)	971.93	R(48)	991.57
P(24)	940.55	R(16)	973.29	R(50)	992.49

9.6 µm band (001 $\rightarrow$ 020 transitions)

Line	$\tilde{\nu}$ (cm^{-1})	Line	$\tilde{\nu}$ (cm^{-1})	Line	$\tilde{\nu}$ (cm^{-1})
P(50)	1016.72	P(20)	1046.85	R(16)	1075.99
P(48)	1018.90	P(18)	1048.66	R(18)	1077.30
P(46)	1021.06	P(16)	1050.44	R(20)	1078.59
P(44)	1023.19	P(14)	1052.20	R(22)	1079.85
P(42)	1025.30	P(12)	1053.92	R(24)	1081.09
P(40)	1027.38	P(10)	1055.63	R(26)	1082.30
P(38)	1029.44	P(8)	1057.30	R(28)	1083.48
P(36)	1031.48	P(6)	1058.95	R(30)	1084.64
P(34)	1033.49	P(4)	1060.57	R(32)	1085.77
P(32)	1035.47	R(4)	1067.54	R(34)	1086.87
P(30)	1037.43	R(6)	1069.01	R(36)	1087.95
P(28)	1039.37	R(8)	1070.46	R(38)	1089.00
P(26)	1041.28	R(10)	1071.88	R(40)	1090.03
P(24)	1043.16	R(12)	1073.28	R(42)	1091.03
P(22)	1045.02	R(14)	1074.65	R(44)	1092.01

ously passed through the discharge tube. This has the added advantage of increasing the population inversion, thereby further improving the efficiency.

A small carbon dioxide laser, with a discharge tube about half a metre in length, may have an efficiency rating as high as 30% and produce a continuous output of 20 W; even a battery-powered hand-held model can produce 8 W cw output. Much higher powers are available from longer tubes; although efficiency drops, outputs in the kilowatt range are obtainable from the largest room-sized devices. Apart from extending the cavity length, another means

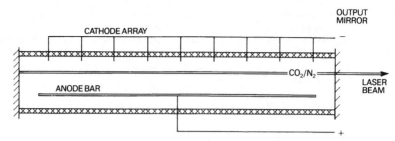

Fig. 2.10 Electrode arrangement in a transverse excitation atmospheric (TEA) carbon dioxide laser

used to increase the output power in such lasers is to increase the pressure of carbon dioxide inside the discharge tube, so increasing the number of molecules available to undergo stimulated emission. In fact, carbon dioxide lasers can be made to operate at or above atmospheric pressure, although in these cases much stronger electric fields must be applied to sustain the discharge. To produce sufficiently strong fields without using dangerously high voltages, it is necessary to apply a potential across rather than along the tube, as illustrated in Fig. 2.10. Such a laser is referred to as a transverse excitation atmospheric (TEA) laser. With higher pressures of around 15 atm, pressure broadening results in a quasi-continuum of emission frequencies, enabling the laser to be continuously tuned over the range 910–1100 cm^{-1}.

In passing, it is worth noting that there is one other very unusual way of pumping a carbon dioxide laser, which does not involve the use of electrical excitation. In the *gas dynamic laser*, a mixture of carbon dioxide and nitrogen is heated and compressed and then injected into a low-pressure laser cavity at supersonic speed. Since the (001) vibration-rotation states have a longer lifetime, the rapid cooling associated with this process depopulates these upper laser levels less rapidly than the lower levels. A population inversion thus ensues, resulting in the usual laser action. Whilst such devices can produce outputs of 100 kW or more, the emission is limited to a few seconds duration, the construction is necessarily unwieldy, and the device has the dubious distinction of being the only laser that is literally noisy.

Carbon dioxide lasers are extensively used in the area of laser-induced chemical reactions, as we shall see in Chap. 5. However, the majority of their *industrial* applications are at present to be found in material processing, such as hole drilling, welding, cutting and surface treatment. Despite the fact that metals in particular tend to be quite reflective in the wavelength region where these lasers operate, the enormous intensities of about 10^{10} Wm^{-2} which CO_2 lasers produce at focus more than compensate. It is also important to note that the total amount of heat transferred to the metal by the laser beam is minimal. Applications of this kind account for the fact that carbon dioxide lasers represent the largest sector of commercial laser sales.

One other important area of application for CO_2 lasers lies in surgical procedures. The cells of which biological tissue is composed largely consist of water and as such can be instantly vapourised by any powerful CO_2 laser beam; moreover, the heat supplied to the surrounding tissue cauterises the wound and prevents the bleeding normally associated with conventional surgery. For these reasons there are an increasing number of operations in which CO_2 laser radiation is proving a more acceptable alternative to the scalpel. The radiation not only provides a very neat method of incision, but in other cases can be used for completely removing large areas of tissue. The development of flexible catheter waveguides for CO_2 laser radiation has increased the scope for applications of this type.

A related gas laser is the carbon monoxide laser. This operates in many respects in a similar way to the carbon dioxide system, with an electrical discharge initially exciting nitrogen molecules. This leads to collisional activation of the carbon monoxide and subsequent laser emission in the range 4.97–8.26 μm. The main difference is that being a diatomic species, CO has only one mode of vibration. In conjunction with suitable pulsing techniques, the line-tunable nature of both CO and CO_2 lasers is such that they are widely used in studies of reaction dynamics based on time-resolved infra-red spectroscopy.

2.4.2
Nitrogen Laser

Another gas laser based on a simple chemically stable molecular species is the nitrogen laser. This laser has three main differences from the carbon dioxide laser. Firstly, it operates on electronic, rather than vibrational transitions. The gas is excited by a high-voltage electrical discharge which populates the $C^3\pi_u$ triplet electronic excited state, and the laser transition is to the lower energy metastable $B^3\pi_g$ state. The second difference results from the fact that the upper laser level, with a lifetime of only 40 ns, has a much shorter lifetime than the lower level, and consequently a population inversion cannot be sustained. The third difference is that essentially all the excited nitrogen molecules undergo radiative decay together over a very short period of time, effectively emptying the cavity of all its energy. This kind of process is known as *super-radiant emission* and is sufficiently powerful that a highly intense pulse is produced without the need for repeatedly passing the light back and forth between end-mirrors. Indeed, the nitrogen laser can be successfully operated without any mirrors, though in practice a mirror is placed at one end of the cavity so as to direct the output.

The configuration of the nitrogen laser is thus similar to that shown in Fig. 2.10, except that the output mirror is absent. The laser therefore automatically operates in a pulsed mode, producing pulses of about 10 ns or less duration at a wavelength of 337.1 nm. Bandwidth is approximately 0.1 nm, and pulse repetition frequency 1–200 Hz. Pulses can tend to have a relatively unstable temporal profile as a result of the low residence time of photons in the laser cavity. Since the laser can produce peak intensities in the 10^{10} Wm^{-2} range, the nitrogen laser is one of the most powerful commercial sources of ultraviolet radiation, and it is frequently used in photochemical studies. It is also a popular choice as a pump for dye lasers, although here it is increasingly being displaced by the more powerful wavelengths available from excimer lasers or harmonics of the Nd:YAG laser.

2.4.3
Chemical Lasers

In all of the lasers we have examined so far, the pumping mechanism used to initiate the population inversion has involved an external source of power. In a chemical laser, by contrast, population inversion is created directly through an exothermic chemical reaction or other chemical means. To be more precise, we can define a chemical laser as one in which irreversible chemical reaction accompanies the laser cycle. This definition is somewhat more restrictive than that used by some other authors, and thereby excludes the iodine and excimer lasers dealt with in the following sections. The concept of a chemical laser is a very appealing one since large amounts of energy can be released by chemical reaction, so that if the laser is efficient in its operation it can produce a very substantial output of energy in the form of light. One of the earliest examples of a chemical laser was the hydrogen chloride laser, whose operation is based on the chemical reaction between hydrogen and chlorine gas, according to the following sequence of reactions:

$$Cl_2 + h\nu_p \rightarrow 2Cl; \tag{2.1}$$

$$Cl + H_2 \rightarrow HCl^\ddagger + H; \tag{2.2}$$

$$H + Cl_2 \rightarrow HCl^\ddagger + Cl; \tag{2.3}$$

where $h\nu_p$ is a pump ultraviolet photon provided by a flashlamp. The appropriate laser transition is provided by the subsequent radiative decay of vibrationally excited hydrogen chloride molecules. Although the free radical propagation reactions (2.2) and (2.3) leading to the production of excited HCl are exothermic, the initiation stage (2.1) requires the input of radiation, and hence an external source of power is still necessary.

A modification of this scheme involving hydrogen fluoride is much more popular and is in commercial production. The only major difference is that in the initiation stage free fluorine radicals are generally produced by an electrical discharge, for example by the electron impact dissociation of a species such as SF_6, which is appreciably less hazardous than the alternative F_2. Oxygen gas is also included in the reaction mixture to scavenge sulphur from the decomposition of SF_6 by reaction to form SO_2. The essential features of the laser are illustrated in Fig. 2.11, where the resonant cavity is anything up to a metre in

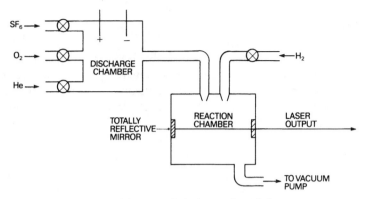

Fig. 2.11 Essential features of a hydrogen fluoride laser

length; the output beam is typically 2–3 mm in diameter and has a divergence of about 2 mrad.

Since only about 1% of the reactant gases are expended on passage through the laser, removal of HF and replenishment of SF_6 and H_2 enables the exhaust gas to be recycled if required. The HF laser produces output in the region 2.6–3.0 μm, consisting of a series of discrete wavelengths associated with rotation-vibration transitions; its deuterium counterpart based on DF transitions emits wavelengths a factor of approximately $2^{1/2}$ longer, i. e. in the range 3.6–4.0 μm. Cw power ranges from a few watts to 150 W. Amongst other research applications, the hydrogen fluoride laser is uniquely suited to absorption-based measurements of HF atmospheric pollution in the neighbourhood of industrial plants. Another example of a chemical laser is afforded by the following scheme utilizing a mixture of nitric oxide, fluorine, deuterium and carbon dioxide gases:

$$NO + F_2 \rightarrow ONF + F; \tag{2.4}$$

$$F + D_2 \rightarrow DF^{\ddagger} + D; \tag{2.5}$$

$$DF^{\ddagger} + CO_2 \rightarrow DF + CO_2^{\ddagger}. \tag{2.6}$$

The result of this sequence of reactions is the collisional transfer of vibrational energy to carbon dioxide molecules, thus creating the conditons under which laser emission can occur exactly as in the carbon dioxide laser discussed earlier. Such a laser essentially has its own built-in power supply and can be turned on simply by opening a valve to allow mixture of the reactant gases. As such, it obviates the need for an external power supply, and thus holds an important advantage over other kinds of laser for many field applications. For higher-power cw chemical lasers, relatively large volumes of reactant gases have to be rapidly mixed at supersonic velocities, and a high degree of spatial homogeneity is required in the mixing region for a stable output.

2.4.4
Iodine Laser

Another type of laser operating on broadly similar principles is the iodine laser. This particular laser refuses to fit squarely into most classification schemes; as we shall see, whilst polyatomic chemistry is involved in its operation, the crucial laser transition actually takes place in free atomic iodine, so that it does not strictly deserve to be classed as a molecular laser. Also, it does not involve irreversible chemistry in the laser cycle, only in side-reactions, so that from this point of view it is equally wrong to class it as a chemical laser. Nonetheless, the iodine laser does share many features in common with the chemical lasers discussed in Sect. 2.4.3, and hence it seems most appropriate to consider it here.

The driving principle involved in the iodine laser, more fully termed the *atomic iodine photodissociation laser*, is the photolysis of iodohydrocarbon or iodofluorocarbon gas by ultraviolet light from a flashlamp. One of the gases typically used for this purpose is 1-iodoheptafluoropropane, C_3F_7I, which is stored in an ampoule and introduced into the silica laser tube at a pressure of between 30 and 300 mbar. The reaction sequence is then as follows:

$$C_3F_7I + h\nu_p \rightarrow C_3F_7 + I^*; \tag{2.7}$$

$$I^* \rightarrow I + h\nu_L; \tag{2.8}$$

$$C_3F_7 + I + M \rightarrow C_3F_7I + M; \tag{2.9}$$

where $h\nu_p$ is a pump photon provided by the flashlamp, and $h\nu_L$ is a laser emission photon; the corresponding energy-level diagram is schematically illustrated in Fig. 2.12. The laser emission stage involves transition between an

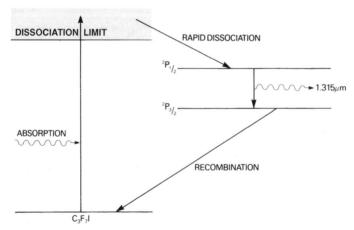

Fig. 2.12 Energetics of the atomic iodine photodissociation laser (not showing fine structure or side-reactions)

excited metastable $^2P_{1/2}$ state and the ground $^2P_{3/2}$ state of atomic iodine; this results in narrow linewidth output at a wavelength of 1.315 μm (7605 cm^{-1}), consisting of six very closely spaced hyperfine components encompassing a range of less than 1 cm^{-1}.

Despite the fact that the sequence of reactions (2.7) to (2.9) in principle represents a repeatable cycle, there are competing irreversible side-reactions which destroy approximately 10% of the active medium during each laser cycle. The main side-reactions are:

$$C_3F_7 + C_3F_7 + M \rightarrow C_6F_{14} + M; \tag{2.10}$$

$$I + I + M \rightarrow I_2 + M. \tag{2.11}$$

Any build-up of molecular iodine through the mechanism of equation (2.11) further reduces the efficiency of the laser process, because it very effectively quenches the upper laser level by the reaction

$$I_2 + I^* \rightarrow I_2^* + I, \tag{2.12}$$

and so diminishes the extent of population inversion. For this reason, the photolysed gas must be exhausted from the laser after emission has taken place, and the tube recharged with fresh gas for the next pulse. The exhaust gas can in fact be recycled provided the molecular iodine is trapped, for example in a low-temperature alkyl iodide solution.

An important advantage of the iodine laser is the fact that the active medium is comparatively cheap and, hence, available in large quantities. Another useful feature is that its emission wavelength lies in a region that offers excellent atmospheric transmission. In the absence of any other pulsing mechanism, the laser typically produces pulses of microsecond duration, and pulse energies of several joules; however, the output is often modified by Q-switching or mode-locking (see Sect. 3.3) to produce pulse trains of nanosecond or sub-nanosecond duration. One application of the laser of particular interest to chemists is the ability to produce a rapid increase of temperature in aqueous media. Since liquid water absorbs very strongly at 1.315 μm with an efficiency of about 30% per centimetre, the iodine laser can induce nanosecond temperature jumps of several degrees Celsius, thus offering a wide range of possibilities for the study of fast chemical and biological solution kinetics. Other uses include lidar and fibre-optical applications.

One variation on the atomic iodine laser is a device known as the *chemical oxygen iodine laser*, in which collisional energy transfer from electronically excited ($^1\Delta$) oxygen molecules to atomic iodine, here formed by dissociation of molecular iodine, populates the $^2P_{1/2}$ state and so leads to the same 1.315 μm laser emission as in (2.8). By employing supersonically flowing gases, multikilowatt cw output powers can be achieved.

2.4.5
Excimer Lasers

The next category of commonly available lasers consists of those in which the active medium is an *exciplex*, or excited diatomic complex. The crucial feature of an exciplex is that only when it is electronically excited does it exist in a bound state with a well-defined potential energy minimum; the ground electronic state generally has no potential energy minimum, or else only a very shallow one. Most examples are found in the inert gas halides such as KrF, whose potential energy curves are illustrated in Fig. 2.13. In principle it is only homonuclear diatomics of the same kind, such as Xe_2, that should be termed *excimers*, but in connection with lasers the designation is now generally applied to heteronuclear exciplexes.

The exciplex is generally formed by chemical reaction between inert gas and halide ions produced by an electrical discharge. For KrF, the exciplex is produced by the following series of reactions:

$$Kr + e \rightarrow Kr^+ + 2e; \qquad (2.13)$$

$$F_2 + e \rightarrow F^- + F; \qquad (2.14)$$

$$F^- + Kr^+ + He \rightarrow KrF^* + He. \qquad (2.15)$$

In the three-body reaction (2.15), the helium simply acts as a buffer. Because the krypton fluoride so produced is electronically excited and has a short life-

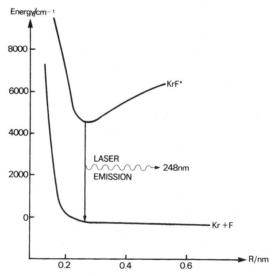

Fig. 2.13 Energetics of a krypton fluoride exciplex laser. Only the electronic states directly involved in laser action are shown

time (about 2.5 ns), it rapidly decays by photon emission to the lower energy state as shown on the diagram. Since this is an unbound state, in which the force between the atoms is always repulsive, the exciplex molecule then immediately dissociates into its constituent atoms. Consequently, this state never attains a large population, and a population inversion therefore exists between it and the higher energy bound exciplex state. The decay transition can thus be stimulated to produce laser emission with a high efficiency, typically around 20%. One noteworthy feature of this particular laser scheme is that it represents a rare example of a genuinely two-level laser.

Although any excimer laser may in a sense be regarded as a type of chemical laser, it is worth noting that the end of a complete cycle of laser transitions generally sees regeneration of the starting materials. Thus in the case of KrF the two processes shown below regenerate krypton and fluorine gas:

$$KrF^* \rightarrow Kr + F + h\nu_L; \tag{2.16}$$

$$F + F \rightarrow F_2. \tag{2.17}$$

Hence the laser can be operated continuously without direct consumption of the active medium, in contrast to the chemical lasers examined earlier. Thus a sealed cavity can be used, in which for the case of KrF the mix of gases is typically 2% krypton, 0.2% fluorine and 97.8% helium, to an overall pressure of 2.5–3.0 atm. Of course, the cavity wall material has to be carefully chosen in view of the highly corrosive halogen gas used. Moreover, since the walls are rapidly poisoned by the gas, it is not possible to use the same laser tube for different halogens.

Excimer lasers are superradiant and produce pulsed radiation with pulse durations of 10–20 ns and pulse repetition frequencies generally in the 1- to 500-Hz range. Pulse energies can be up to 1 J, with peak pulse power in the megawatt region and average power between 20 and 100 W. The emission wavelengths of commercially available systems are as follows: F_2 157 nm; ArF 193 nm; KrCl 222 nm; KrF 248 nm; XeCl 308 nm; XeF 351 and 353 nm. These short ultraviolet wavelengths lie in a region of the electromagnetic spectrum absorbed by a very wide range of materials, and the photoabsorption process often results in a rupture of chemical bonds. Often such absorption also leads to a degree of sample vapourisation; this is a process known as laser *ablation*. This feature, coupled with excellent beam quality, good spatial control and high power levels usually produced, make excimer lasers the logical choice for a diverse range of materials processing, surgical and photochemical applications (for the latter, see Chap. 5).

Excimer laser treatment normally results in sharply localised surface vapourisation, without the thermal expansion effects on surrounding material which would often result from the use of an infra-red laser. Particularly useful in this connection is the facility to optically sculpt the beam to produce radiation with an essentially tabular beam profile – see Fig. 1.11c and d, p. 18. Such lasers are, for example, unparalleled in precision metal cutting applications, offering submicron bore diameters in optimum cases. The etching and marking of high density materials provides another illustration, which opens up a number of possibilities for the security marking of valuables such as diamonds – a similar technique is already routinely used to trademark car mirrors. Lower power levels can be employed in surface cleaning, as for example with objets d'art. It has also been demonstrated that there are significant surgical applications; tissue irradiated with excimer laser wavelengths undergoes molecular fragmentation and volatilisation without thermal damage. An increasingly popular ophthalmic procedure based on this property is *photorefractive keratectomy*, the corneal sculpting treatment for nearsightedness. In conjunction with an optical fibre catheter, such lasers have also proved highly effective at removing plaque from clogged arteries.

2.5
Dye Lasers

The last category of commonly available lasers consists of dye lasers. These are radically different from each of the types of laser we have discussed so far. All of the differences can be traced back to the unusual nature of the active medium, which is a solution of an organic dye. A wide range of over 200 dyes can be used for this purpose; the only general requirements are an absorption band in the visible spectrum and a broad fluorescence spectrum. The kind of compounds which satisfy these criteria best consist of comparatively large polyatomic molecules with extensive electron delocalisation; the most widely used example is the dye commonly known as Rhodamine 6G ($C_{28}H_{31}N_2O_3Cl$), a chloride of the cation whose structure is shown in Fig. 2.14. With 64 atoms, this species has 186, (3N-6), distinct modes of vibration. In solution, the corresponding energy levels are of course broadened due to the strong molecular

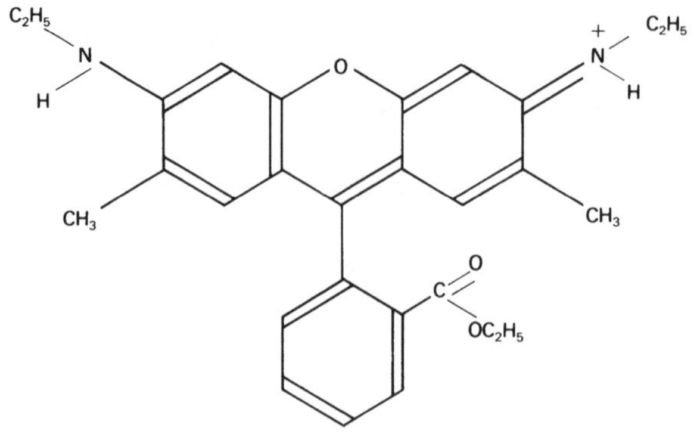

Fig. 2.14 One of the resonance structures of the Rhodamine 6G cation

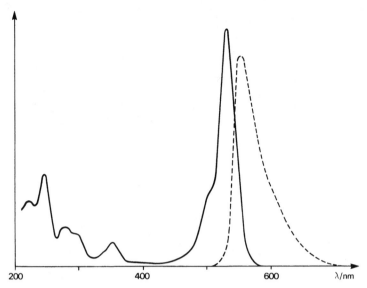

Fig. 2.15 Solution spectra of Rhodamine 6G in ethanol. The solid curve shows
the absorption, and the dotted curve the fluorescence at longer wavelength

interactions of the liquid state, and they overlap to such an extent that an energy continuum is formed for each electronic state. This results in broadband
absorption and fluorescence, as illustrated by the Rhodamine 6G spectra in
Fig. 2.15.

Generally speaking, the absorption of visible light by polyatomic dyes first
of all results in a transition from the ground singlet state S_0 to the energy continuum belonging to the first excited singlet state S_1. The singlet designation
arises from the fact that a state with no unpaired electron spin is non-degenerate (i.e. $2S + 1 = 1$ if $S = 0$). This is immediately followed by a rapid radiationless decay to the lowest energy level within the S_1 continuum, as illustrated
in Fig. 2.16; in the case of Rhodamine 6G, this process is known to be complete
within 20 ps of the initial excitation. Fluorescent emission then results in a
downward transition to levels within the S_0 continuum, followed by further
radiationless decay. It is the fluorescent emission process which can be made
the basis of laser action, provided a population inversion is set up between the
upper and lower levels involved in the transition; we are therefore essentially
dealing with a four-level laser (compare Fig. 1.8, p. 15). Clearly, since the
emitted photons have less energy, the fluorescence must always occur at longer
wavelengths than the initial excitation.

However, as the dotted arrows on the diagram show, there are usually several other processes going on which complicate the picture. One of the most
important of these competing transitions is known as *intersystem crossing*,
since the molecule changes over from one of the system of singlet states to a

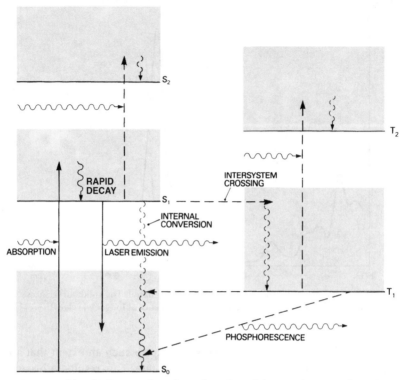

Fig. 2.16 Jablonski diagram for a laser dye. The solid vertical arrows show the transitions involved in laser action, and the dotted arrows illustrate some of the competing transitions

triplet state, in which two electron spins are parallel ($2S + 1 = 3$ if $S = 1$). Such singlet-to-triplet transitions which are formally forbidden nonetheless take place at a small but significant rate. The T_1 state is to a small extent depopulated by the comparatively slow process of phosphorescence, which is also spin-forbidden and returns the molecule to its ground electronic state S_0. It is also depopulated through non-radiative intersystem crossing to S_0 and through further absorption of radiation which populates higher triplet states. In addition to all these processes, a molecule in the singlet state S_1 can also undergo radiationless *internal conversion* to the S_0 state or absorb further radiation and undergo a transition to a higher singlet state. Together, these processes contribute to a depopulation of the upper S_1 laser level, an increase in the population of the lower S_0 laser levels, and a reduction in the output intensity, all of which contribute to a decrease of laser efficiency.

Quenching of the excited states through interaction with other molecules also occurs, however: and this effect is particularly important if the dye solution contains dissolved oxygen. In some cases triplet quenchers such as dimethylsulfoxide (DMSO) are specially added to the dye solution to increase

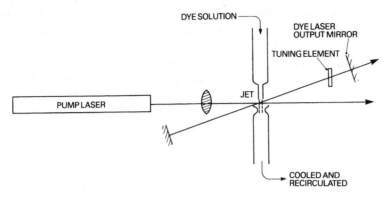

Fig. 2.17 Schematic diagram of a laser-pumped dye laser

output power by repopulating singlet states. The photochemical and thermal stability of the dyes used in dye lasers is another clearly important factor. The heat produced by the radiationless decay transitions can very rapidly degrade a dye, and for this reason it is common practice to continuously circulate the dye solution, to allow cooling to take place. A common setup is shown in Fig. 2.17. The pump radiation from a flashlamp or a primary laser source with emission in the visible or near-ultraviolet range is focussed at a point traversed by a jet of dye solution, which typically has a concentration in the range 10^{-2}–10^{-4} mol 1^{-1}. The dye solvent is usually based on ethylene glycol, which provides the viscosity needed to maintain an optically flat jet stream.

The fluorescent emission from the dye jet is stimulated by placing two parallel cavity-end mirrors on either side of the jet. However, it is at this stage that the unique properties of the dye laser become evident. Since the fluorescence occurs over a range of wavelengths, monochromatic laser emission can only be obtained by the use of additional dispersive optics such as a diffraction grating or etalon within the cavity. However, rotation of this element obviously changes the wavelength amplified within the dye laser, so that *tunable emission* is obtained. A dye laser based on a solution of Rhodamine 6G in methanol, for example, is continuously tunable over the range 570–660 nm. The full range of commercially available dye lasers provides complete coverage of the range 200 nm–1 μm: typical tuning curves for some of the more important dyes are given in Fig. 3.5 (Sect. 3.2.1). A thoroughly comprehensive listing of laser dyes and tuning ranges can be found in the book by Maeda.[2]

The efficiency of a dye laser is often around 5%, and the output power depends principally on the source of the pumping radiation. For cw output, the usual pump is an inert gas ion laser; other commonly used laser pumps such as the nitrogen, excimer or solid-state transition-metal ion lasers, or else xenon flashlamps, generally result in pulsed output. Use of a dye laser as a means of

2) Maeda M (1984) 'Laser dyes' Academic, New York

laser frequency conversion is discussed further in Sect. 3.2. The highest powers are obtained by pumping with harmonics of a Nd laser. Cw dye lasers produce emission with linewidth in the range 10–20 GHz (around 0.5 cm^{-1}), although with suitable optics this can be reduced to about 1 GHz. The combination of narrow linewidth, good frequency stability and tunability is particularly attractive for many spectroscopic applications. One disadvantage of the dye laser is that it tends to have rather poorer amplitude stability than a gas laser, so that indirect spectroscopic methods such as fluorescence or the photoacoustic effect are often most appropriate.

One interesting variation on the dye laser concept is the *ring dye laser*, in which the laser radiation travels around a cyclic route between a series of mirrors, rather than simply back and forth between two. When both clockwise and anticlockwise travelling waves are present, their frequencies are normally identical. However, any rotation of the laser itself results in a small difference between these two frequencies, and detection of this difference can be used as the basis for very accurate measurement of the rotation; this is the principle behind the ring laser gyroscope. Alternatively, an optical element which behaves somewhat like an optical counterpart to an electronic diode can be placed within the cavity to select one particular direction of propagation (clockwise or anticlockwise). In this case the ring laser has an emission linewidth typically at least a factor of ten smaller than a conventional dye, but in state of the art actively stabilised cases the linewidth may be as small as 10^{-15} cm^{-1}.

2.6
Free-Electron Laser

The last type of laser we shall consider is one that is radically different from any of those mentioned so far. This is a laser in which the active medium consists purely and simply of a beam of free electrons, and the optical transitions on which laser action is based result from the acceleration and deceleration of these electrons in a magnetic field. One of the most common experimental arrangements is illustrated in Fig. 2.18 and involves passing a beam of very high-energy electrons from an accelerator between the poles of a series of regularly spaced magnets of alternating polarity. The electrons typically need to have energies in the 10^7–10^8 MeV range, with the magnet spacing a few centimetres. The result is that the electrons are repeatedly accelerated and decelerated in a direction perpendicular to their direction of travel, resulting in an oscillatory path, as shown. For this reason, the magnets are generally referred to as *wiggler magnets*. The effect of this process is to produce emission of *bremsstrahlung* radiation along the axis of the laser, which is then trapped between parallel mirrors and stimulates further emission in the usual way.

The wavelength of light emitted by the free-electron laser is determined by the energy of the electrons and the spacing of the magnets. High-energy elec-

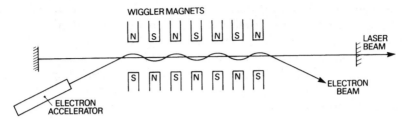

Fig. 2.18 Configuration of a free-electron laser. The electron accelerator is in fact a massive piece of equipment on a far different scale to the ancillary equipment associated with most lasers

trons travel at an appreciable fraction of the speed of light and can be characterised by a parameter f, denoting the ratio of their relativistic total energy to their rest-mass energy. If we denote the magnet separation by d, then the laser emission wavelength is given by the simple formula:

$$\lambda = d/2f^2. \tag{2.18}$$

Since the energy of the electrons emerging from the accelerator can be continuously varied, the result is once again a laser with a tuning capability.

In contrast to the dye laser, however, there is in principle no limit to the range of tuning right across the infra-red, visible and ultraviolet regions of the electromagnetic spectrum. Moreover, this kind of laser has been shown to produce high average powers and to be capable of a reasonable efficiency. For example, a setup in the Naval Research Laboratory in the USA can produce 75-MW pulses of 4-mm radiation with an efficiency of about 6%. The efficiency in the visible and ultraviolet is, nonetheless, generally much lower, and a great deal of research effort is currently being directed towards this problem. For many other reasons, not least the high power requirement and large bulk of a suitable electron accelerator, the free-electron laser is not in commercial production and is likely to remain in the province of highly specialised research equipment for the time being. Nevertheless, for applications which require tunable radiation at very high power levels, the free-electron laser will prove hard to beat.

2.7
Questions

1. The ruby laser emits red light. Why does this simple fact immediately show that its energetics involve more than two energy levels?
2. The crystal employed in a ruby laser (operational wavelength 694 nm) typically contains approximately 2×10^{19} Cr^{3+} ions per cubic centimetre. Under optimum Q-switched conditions, almost all of the ions in the laser rod can be promoted to the upper laser level just before emission occurs. What is the

maximum possible energy output in a single pulse from a ruby rod 7.5 cm long and 1 cm in diameter? What is the peak power output if the pulse duration is 100 ns? Roughly how many electric light bulbs would you need to produce an equivalent power? ($h = 6.63 \times 10^{-34}$ J s: $c = 3.00 \times 10^8$ m s^{-1}).

3. Contrast the ways in which metals can be used as sources of light emission in lasers and conventional light sources and outline the main reasons why the light emitted by two such sources differs in its properties.

4. The rotation-vibration transitions on which the emission of light from a carbon dioxide laser is based give rise to output at about 50 closely spaced but discrete frequencies in the 10.6 μm band ranging from 907.78 cm^{-1} to 992.49 cm^{-1}, and another 45 in the 9.6 μm band ranging from 1016.72 cm^{-1} to 1092.01 cm^{-1}. However, in a high pressure carbon dioxide laser, emission is continuously tunable over the entire range. Why?

5. Estimate the maximum rate at which a 1 kW cw carbon dioxide laser can cut along 3 mm aluminium sheet, assuming a cut width of 1 mm and 80% conduction losses. The ambient temperature may be taken as 20°C. (Specific heat of aluminium = 878 J kg^{-1} K^{-1}; melting point = 932 K; latent heat of fusion = 3.95×10^5 J kg^{-1}; density = 2.70×10^3 kg m^{-3}).

6. The hydrogen fluoride laser operates on vibration-rotation levels and produces output in the region 2.6–3.0 μm. What kind of wavelength range would you expect the deuterium fluoride laser to cover?

7. With reference to the principal reactions and side-reactions involved in the atomic iodine photodissociation laser, explain why it is often necessary to recycle the gases through a low-temperature alkyl iodide solution. Explain also why this laser is particularly suited to the study of fast reactions in aqueous solution.

8. The thermal and photochemical stability of laser dyes is closely related to their long-wavelength limit of absorption, and it is generally considered undesirable for viable dyes to display absorption too far into the near-infra-red. One such dye absorbing in the near infra-red has a low-lying excited singlet state and a metastable triplet state somewhat lower in energy. Dye molecules can reach this highly reactive state by thermal activation and often subsequently decompose by reaction with solvent molecules, dissolved oxygen, impurities or other dye molecules. The decomposition is of the pseudo-first-order type with a reaction rate constant given by $k = A \exp(-E_a/RT)$, where E_a is the activation energy and the pre-exponential factor $A \approx 10^{12}$ s^{-1}. Taking the minimum practical half-life of the dye as 1 day, estimate the maximum wavelength of absorption that will be tolerable. ($R = 8.31$ J K^{-1} mol^{-1}, $L = 6.02 \times 10^{23}$ mol^{-1}, $h = 6.63 \times 10^{-34}$ J s).

CHAPTER 3

Laser Instrumentation in Chemistry

I tune the Instrument here at the dore,
And what I must doe then, thinke now before
'Hymne to God my God, in my Sicknesse', John Donne

In the first two chapters of this book, we have considered the chemical and physical principles underlying the operation of various types of laser, and the characteristic properties of the light which they emit. In the remainder of the book, we shall be concerned with chemical applications of lasers, paying particular attention to the ways in which each application has been developed to take the fullest advantage of the unique properties of laser light.

Laser instrumentation is now widely used to probe systems of chemical interest. These systems may be chemically stable, or they may be in the process of chemical reaction, but in either case the laser can be utilised as a powerful analytical device. There are numerous and widely diverse ways in which the laser is now used in chemical laboratories, and this chapter provides an introduction to some of the general instrumental principles concerned. By far the most widely used methods are spectroscopic in nature, and a more detailed discussion of the enormously varied chemical applications in this area is reserved for the following chapter. The quite distinct area of chemistry in which the laser is used to promote chemical reaction is dealt with in Chap. 5. Other chemical applications which fall outside the province of Chaps. 4 and 5 will be examined here.

We begin with a discussion of some of the ways in which laser output can be modified. We have seen that most of the important emission characteristics of any particular type of laser are determined by the energy levels and kinetics associated with the cycle of laser transitions in the active medium; the physical properties of the laser cavity and the nature of the pump are also involved, as discussed earlier. However, with any laser there are certain applications for which the optimum specifications are somewhat different, and it is therefore desirable to tailor the laser output in certain ways. Three particularly important and widely used modifications involve polarisation modification, frequency conversion, and pulsing.

3.1
Polarising Optics

Every photon has an associated electric and magnetic field which oscillate perpendicularly to its direction of propagation. The polarisation of a monochromatic beam of photons is one measure of the extent of correlation between these oscillations. Some important types of beam polarisation are illustrated in Fig. 3.1. If the resultant of the photon oscillation lies in a single plane, as in (a), then we have *plane polarisation*, also known as *linear polarisation*. If the resultant describes a helix about the direction of propagation, as in (b) and (c), then we have left- or right-handed *circular polarisation*. Finally, there is the case of *elliptical polarisation*, illustrated by (d), which is essentially an intermediate between (a) and (c). If there is no correlation between the photons, then the state of the beam is regarded as *unpolarised* and may be represented by two orthogonal plane polarisations with no phase relationship between them.

Laser operation does not by itself automatically produce polarised light. The simple diagram of Fig. 1.4 shows a cavity mode which is plane polarised in the plane of the diagram; however, plane polarisation perpendicular to the plane of the diagram is equally possible. In the absence of any discriminatory optics both will be present in the cavity and will lead to an unpolarised laser output. For a number of laser applications, however, it is desirable to obtain polarised emission, and it is usual to incorporate optics into the laser cavity to accomplish this. The standard method is to use a *Brewster angle window*. This is a parallel-sided piece of glass set at the Brewster angle, defined as

$$\theta_B = \tan^{-1} n, \tag{3.1}$$

where n is the refractive index of the glass for the appropriate wavelength. The Brewster angle is the angle at which reflected light is completely plane polarised. When such elements are used for the end-windows of a gas laser tube, for example, then only one selected polarisation experiences gain and is amplified within the cavity. Polarised emission thus ensues as illustrated in Fig. 3.2. An alternative method of achieving the same result is to make use of a birefringent crystal such as calcite ($CaCO_3$), in which orthogonal polarisations propagate in slightly different directions and can thus be separated.

Calcite is a good example of an optically anisotropic material, i. e. one whose optical properties such as refractive index are not the same in all directions. Crystals of this type have a number of other important uses in controlling laser polarisation. Mostly, these stem from the fact that in such *birefringent* media different polarisations propagate at slightly different velocities. By passing *polarised* light in a suitable direction through such a crystal, the birefringence effects a change in the polarisation state, the extent of which depends on the distance the beam travels through the crystal. The changing state of polarisation of an initially plane polarised beam as it traverses an optically

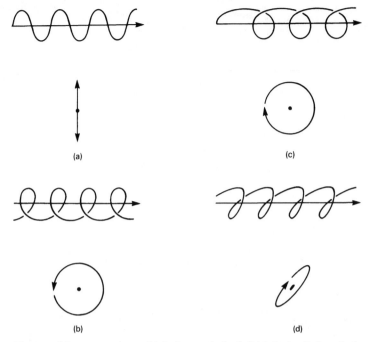

Fig. 3.1a-d Representations of (**a**) plane polarised, (**b**) left circularly polarised, (**c**) right circularly polarised, and (**d**) elliptically polarised light. The lower diagrams show the apparent motion of the electric field vector as the beam travels towards the observer

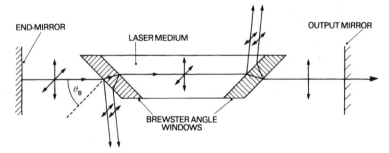

Fig. 3.2 The use of Brewster angle windows in a gas laser. Each window reflection results in loss of the same polarisation component from the cavity. Radiation which is unpolarised as it leaves the left-hand end-mirror thus becomes nearly plane polarised before it reaches the output mirror. The windows thus enable only one polarisation component to be amplified as light passes back and forth between the end-mirrors

anisotropic crystal is shown at the top of Fig. 3.3. As shown beneath, with a slice of crystal of the correct thickness d, plane polarised light thus emerges circularly polarised (a), or vice versa (b), and with a crystal of twice the thick-

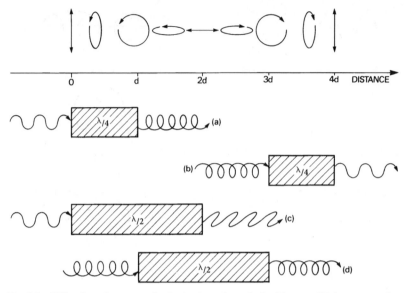

Fig. 3.3a-d The changing state of polarisation of a polarised beam of light propagating through an optically anisotropic medium; (**a**) and (**b**) represent the use of a quarter-wave plate; (**c**) and (**d**) the use of a half-wave plate; see text for details

ness, plane polarised light emerges with a rotated plane of polarisation (c), or circularly polarised light reverses its handedness (d). The two optical elements based on these principles are known as a *quarter-wave plate* and a *half-wave plate,* respectively.

One other type of polarising optic which is particularly useful in many laser applications is the *Pockels cell.* This is a cell consisting of a crystalline material such as potassium dihydrogen phosphate (KDP) which exhibits the Pockels effect, essentially a proportional change in refractive index on application of an electric field. (There is also in general a change in refractive index proportional to the *square* of the electric field; this is known as the *Kerr effect.*) A Pockels cell can once again be used to rotate the plane of polarisation of light passing through it or to reverse the handedness of circularly polarised light. However, its virtue is that because the Pockels effect only takes place whilst the electric field is applied, modulation of the field at a suitable frequency results in a corresponding modulation of the beam polarisation. This is a particularly useful way of increasing detection sensitivity in many techniques based on optical activity, such as polarimetry (Sect. 3.7) and circular differential Raman scattering (Sect. 4.5.6).

3.2
Frequency Conversion

We now turn to another more important way in which laser output can be modified. Since for a given laser the output frequencies are determined by the nature of the lasing material, it is often helpful to be able to convert the output to a different frequency more suited to a particular application. The two most widely adopted methods of frequency conversion involve dye lasers and frequency-doubling crystals.

3.2.1
Dye Laser Conversion

The principle behind the operation of the dye laser has already been discussed in some detail in Sect. 2.5. Its use as a frequency-conversion device follows a similar line, using a primary laser source (often an inert gas or nitrogen laser) rather than the broadband emission from a flashlamp as the pump. Thus the excitation is often essentially monochromatic, and the excitation and emission curves are as shown in Fig. 3.4. The most important feature of this method of frequency conversion is that the dye laser output has to occur at a lower frequency, and hence a *longer wavelength*, than the pump laser. Nevertheless, the exact range of output wavelengths depends on the choice of dye, and the facility for tuning within this range provides a very useful method of obtaining

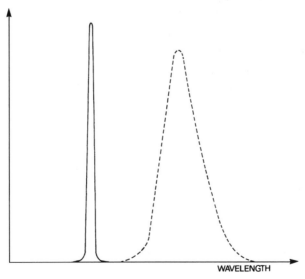

Fig. 3.4 Incident radiation (solid curve) and emission (dotted curve) in a dye laser pumped by a single line from an ion laser

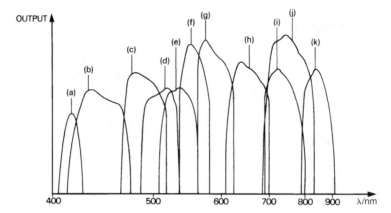

Fig. 3.5 Tuning curves for a dye laser pumped by various lines from a krypton/argon laser. The dye and pump wavelength for each curve are given in Table 3.1. (Reproduced by kind permission of Coherent Radiation Ltd.)

Table. 3.1 Dyes and pump wavelengths for the tuning curves of Fig. 3.5

Curve	Dye	Pump laser; wavelength λ/nm
(a)	Stilbene 1	Argon; 333.6, 351.1, 363.8
(b)	Stilbene 3	Argon; 333.6, 351.1, 363.8
(c)	Coumarin 102	Krypton; 406.7, 413.1, 415.4
(d)	Coumarin 30	Krypton; 406.7, 413.1, 415.4
(e)	Coumarin 6[a]	Argon; 488.0
(f)	Rhodamine 110	Argon; 514.5
(g)	Rhodamine 6G	Argon; 514.5
(h)	DCM[b]	Argon; 488.0
(i)	Pyridine 2	Argon; 488.0, 496.5, 501.7, 514.5
(j)	Rhodamine 700	Krypton; 647.1, 657.0, 676.5
(k)	Styryl 9M	Argon; 514.5

[a] With cyclo-octatetraene and 9-methylanthracene;
[b] 4-dicyanomethylene 2-methyl 6-(p-dimethylaminostyrol) 4H-pyran

laser emission at a chosen wavelength. For example, by using various lines from a krypton/argon ion laser as pump, coherent radiation can be obtained right across and well beyond the visible spectrum, as shown in Fig. 3.5. Conversion efficiencies are generally somewhat low; 5%–10% is not unusual, although with certain dyes such as Rhodamine 6G, figures of 20% or more are possible.

3.2.2
Non-linear Optics

The second commonly used method of frequency conversion, known as *frequency doubling*, is the best known example of a *non-linear optical* process.

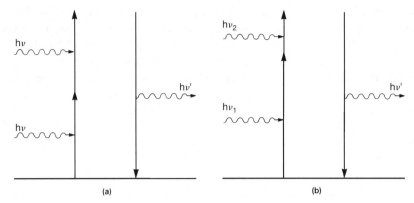

Fig. 3.6a and b Energetics of (**a**) frequency doubling, where the emitted photon has frequency $\nu' = 2\nu$, and (**b**) frequency mixing, where the emitted photon has frequency $\nu' = \nu_1 + \nu_2$

The latter term refers to a wide range of frequency-conversion effects that are non-linearly dependent on laser intensity, such that their conversion efficiencies generally improve when laser power is increased. The energetics of frequency doubling, or to use the physicists' term, *second harmonic generation*, are illustrated in Fig. 3.6.

Two photons of laser light with frequency ν are taken up by a substance in its ground state, and a single photon of frequency ν' is emitted through a transition back to the ground state. Note that the entire process is a concerted one, and there is no intermediate excited state with any measurable lifetime; hence, the energy–time uncertainty principle allows the process to take place regardless of whether there are energy levels of the substance at energies $h\nu$ or $2h\nu$ above the ground state. Indeed, it is usually better if there are not, since the presence of such levels can lead to competing absorption processes. Symmetry considerations reveal that the frequency-doubling process can take place only in media that lack a centre of symmetry – generally a non-centrosymmetric crystal is employed. Though liquids can very weakly generate second harmonics where their bulk symmetry is broken, a feature which, incidentally, can prove useful for the characterisation of liquid surfaces, they are of no use for efficient frequency conversion.

It is important to note that the process of harmonic generation is *coherent*, so that with laser input the emission also has typically laser-like characteristics. In order for coherent output to be obtained, however, it is necessary to arrange for the harmonic conversion process to conserve photon momentum. Generally, photons carry a momentum $hk/2\pi$ in the direction of propagation, where k is the *wave-vector* of magnitude $k = 2\pi\nu/c' = 2\pi/\lambda$, and n is the refractive index of the medium for the appropriate frequency. It can readily be seen that for harmonic emission in the forward direction, momentum is only conserved if the harmonic photons have wave-vector $k' = 2k$, where k is the wave-vector of the laser photons. This in turn leads to a requirement for the refractive indices of the pump beam and the harmonic to be equal. Fulfilment of this condition, known as *index matching* or *wave-vector matching*, can only be accomplished in optically *anisotropic* solids where refractive index is dependent on the direction of propagation and polarisation of light (the phenomenon known as *birefringence*). Here it is often possible to obtain index matching by careful orientation of the crystal. Since refractive indices are often also very heat-sensitive, fine control of the temperature is normally required, and the crystal is accordingly often mounted inside a temperature-controlled heating cell. Either the crystal orientation or its temperature can then be varied for any particular incident wavelength to obtain maximum efficiency. Conversion efficiencies are often as high as 20%–30% and under optimum conditions higher still. Frequency doubling is particularly useful for generating powerful visible (532 nm) radiation using a Nd laser pump.

Two of the most widely used crystals are KDP (KH_2PO_4) and lithium niobate (Li_3NbO_4). For tunable UV generation the excellent characteristics of β-barium borate, with a transparency range of 190–3500 nm, represent a very attractive, if somewhat expensive, alternative. Whilst the majority of crystals used for frequency-doubling and other non-linear processes are crystals of inorganic salts, there is wide interest in the development of non-linear optical devices based on organic solids.[3] One of the most studied is urea, although other compounds such as MNA (2-methyl-4-nitroaniline) and DLAP (deuterated l-arginine phosphate) display more impressive non-linear characteristics. Development work in this area requires consideration not only of crystal structure and non-linear optical properties, but also factors such as ease of crystal growth, mechanical strength, synthetic accessibility and cost.

The generally high conversion efficiency of the frequency-doubling process makes it possible to use a series of crystals to produce 4×, 8×, 16 × ..., the pump frequency, and so to obtain coherent radiation at very short wavelengths. More commonly, one doubling crystal is coupled with a dye laser so as to obtain tunable emission over a range of wavelengths below the operating wavelength of the pump. Another technique which has passed from the research stage to incorporation in a commercial device is *third harmonic generation*, which can take place in a variety of either centrosymmetric or non-centrosymmetric crystals, and is now used to generate 355-nm radiation from Nd:YAG pump lasers.

There are other means of frequency conversion also worth mentioning. One is *frequency mixing*, a variation of the frequency-doubling process in which two beams of laser radiation with different frequencies are coupled in a suitable crystal (again, one which is of non-centrosymmetric structure) to produce output at the sum frequency, as illustrated in Fig. 3.6 b. Since wavelength is

3) Chemla DS, Zyss J (1987) Non-linear optical properties of organic molecules and crystals, vols 1,2. Academic, Orlando

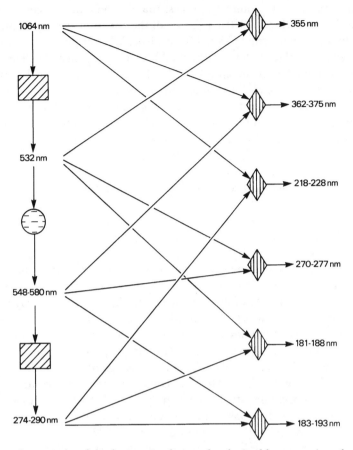

Fig. 3.7 Some of the frequencies that can be obtained by conversion of the 1.064 µm output from a Nd:YAG laser. The rectangles represent frequency doubling, the circle represents dye laser conversion (the range of wavelengths given corresponds to use of Rhodamine 6G solution), and the diamonds represent frequency mixing. Raman shifting, not included here, can also be employed (see text)

inversely proportional to frequency, the emitted wavelength is given by the formula

$$1/\lambda' = 1/\lambda_1 + 1/\lambda_2, \tag{3.2}$$

and hence is always shorter than either of the input wavelengths. Together with dye lasers and frequency-doubling, this method of frequency conversion enables a single laser to be employed for the production of coherent radiation at a variety of wavelengths, as shown in Fig. 3.7. A closely related application of non-linear optics concerns the *optical parametric oscillator*. This again involves frequency mixing in a suitable crystal such as lithium niobate, but is here used to generate tunable radiation by a process that is essentially the converse of Eq. (3.2), in the sense that the left-hand side of the equation refers to input rather than output. Often the fundamental from a Nd:YAG laser pumps the non-linear crystal, which is placed along with suitable frequency-selective elements in a secondary cavity (as in the case of a dye laser). The result is that two new frequencies (the signal and idler frequency) are formed whose sum equals that of the pump beam. Wide tunability and high conversion efficiencies can be obtained in the infra-red by this method. The same principles can be applied in the visible and ultraviolet by adoption of a different pump such as an excimer laser or one of the harmonics of a Nd:YAG laser. For example, pumping beta-barium borate crystal with fourth-harmonic 266-nm radiation enables tunable output to be obtained anywhere from 300 nm to 2.5 μm.

3.2.3
Raman Shifting

One other alternative for frequency conversion is Raman shifting, in which the *stimulated Raman effect* is employed for conversion to either shorter or longer wavelength. The principle of the Raman effect, which is more widely used in a spectroscopic context, will be discussed in detail in Chap. 4. For the present, we simply note that it enables laser frequency to be modified by discrete increments (Stokes and anti-Stokes shifts). For the purpose of frequency conversion, the process is usually accomplished by passage of the laser light through a suitable crystal or a stainless steel cell containing gas at a pressure of several atmospheres. Conversion efficiency for the principal Stokes shift to longer wavelength can be as high as 35%. The nature of the crystal or gas determines the frequency increment; $Ba(NO_3)_2$ gives a shift of 1047 cm^{-1}, whilst the most commonly used gases H_2, D_2 and CH_4 produce shifts of 4155, 2987 and 2917 cm^{-1} respectively. Atomic vapours can also be useful for this purpose; for example 455-nm radiation in the blue-green region best suited to submarine-satellite optical communication can be obtained from the 308-nm output of a xenon chloride excimer laser by Raman conversion in lead vapour.

3.3
Pulsing Techniques

In Chap. 2, we encountered several lasers which can naturally operate on a continuous-wave (cw) basis, whilst others are inherently pulsed, depending on kinetics considerations, as discussed in Sect. 1.4.3. Nonetheless, there are several reasons why it is common practice to make use of a pulsing device, either to convert a cw laser to pulsed output or else to shorten the duration of the pulses emitted by a naturally pulsed laser. One motive is to obtain high peak intensities, since if in any period the emission energy can be stored up and then emitted over a much shorter period of time, the result is an increase in the instantaneous irradiance. Another motive is to produce pulses of very short duration in order to make measurements of processes which occur on the same kind of timescale. There are three widely used pulsing methods, which we now consider in detail.

3.3.1
Cavity Dumping

As the name implies, cavity dumping refers to a method of rapidly emptying the laser cavity of the energy stored within it. The simplest method of accomplishing this with a cw laser would be to have both of the end-mirrors fully reflective, but with a third coupling mirror able to be switched into the beam to reflect light out of the cavity, as shown in Fig. 3.8a. This configuration would deliver a single pulse of light, terminating laser action until the coupling mirror was again switched out of the beam. In practice, an acousto-optic modulator is usually employed, as illustrated in Fig. 3.8b. This double-pass device is driven by a radio-frequency electric field and generates an acoustic wave which produces off-axis diffraction of the laser beam. The cavity-dumped output of the laser in this case consists of an essentially sinusoidal temporal profile as shown in Fig. 3.8c; the frequency of oscillation is twice the acoustic frequency of the cavity dumper, typically of the order of megahertz. This method of pulsing is often used in conjunction with mode-locking, as we shall see in Sect. 3.3.3.

3.3.2
Q-Switching

In physical terms, the reasoning behind this method of achieving pulsed operation is as follows. If a shutter or some other device is placed within the laser cavity so as to increase the loss per round-trip of the radiation, a sizeable population inversion can be established in the active medium without any appreciable amount of stimulated emission taking place. If the shutter is then opened up so that the cavity can properly act as a resonator, the energy stored by the medium can be released in a single pulse of highly intense light. The term *Q-switching* refers to the fact that such a method essentially involves first reducing and then swiftly increasing the quality, or Q-factor, of the laser. In Eq. (1.31), Q was expressed in terms of the resonant frequency and linewidth of the emission. However, that equation is formally derived from another defining equation which is as follows:

$$Q = 2\pi \times \text{energy stored in cavity/energy loss per optical cycle,} \qquad (3.3)$$

in which the optical cycle length is simply the inverse of the resonant frequency. Q-switching thus represents the effect of suddenly reducing the rate of energy loss within the laser cavity. In practice, the pumping rate has to ex-

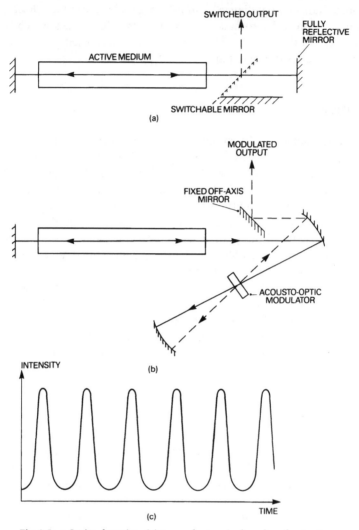

Fig. 3.8a-c Cavity dumping, (**a**) to produce a single pulse of output, and (**b**) for a radio-frequency modulated output with temporal profile as in (**c**)

ceed the rate of spontaneous decay in order to build up the required population inversion, and the time taken for Q-switching has to be sufficiently short to produce a single pulse.

There are three commonly used methods of Q-switching a laser. The first and simplest method involves replacing one of the cavity end-mirrors with one which revolves at a high speed (typically 500 revolutions per second), as illustrated in Fig. 3.9 a. In this situation, cavity losses are kept high except for

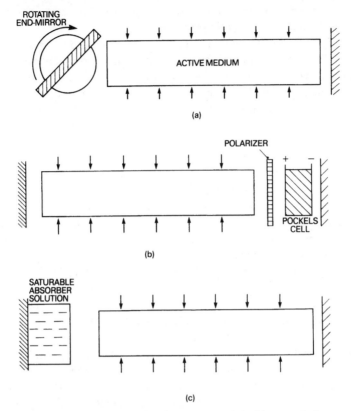

Fig. 3.9a-c Three schemes for obtaining Q-switched laser emission;
(**a**) using a rotating end-mirror, (**b**) an electro-optic switch, and (**c**) a
saturable absorber cell

the brief periods when this mirror is virtually parallel to the static output mirror. The pulse repetition rate is determined by constraints on the rate of pumping, however.

In the second method, there is a genuine shutter action within the cavity, but one which is usually based on electro-optical rather than mechanical principles. One of the devices commonly used is the Pockels cell described in Sect. 3.1. In this case the cell is designed such that when a suitable potential difference is applied along it, plane polarised light traversing the cell and reflecting back through it suffers an overall 90° rotation of the polarisation plane. Hence, with both a polariser and a Pockels cell within the laser cavity, as shown in Fig. 3.9 b, there is an effective shutter action when the voltage is applied, since the polariser cuts out all the light with rotated plane of polarisation. By switching the voltage off, the cavity again becomes effectively transparent, and Q-switched emission can take place. With this method, the switching time is synchronised with the pumping and is typically on the nanosecond scale. There are several variations on this theme based on other electro-optic, magneto-optic and acousto-optic principles.

The third method of Q-switching involves use a solution of a *saturable absorber* dye. This is a compound, which must display strong absorption at the laser emission wavelength, for which the rate of absorption nonetheless *decreases* with intensity; an example used for the ruby laser is cryptocyanine. In a certain sense, then, such a dye undergoes a reversible bleaching process due to depopulation of the ground state. Thus if a cell containing a saturable absorber solution is placed within the laser cavity as shown in Fig. 3.9c, then in the initial stages of pumping the strong absorption effectively renders the cell opaque (in other words, it has negligible transmittance), and losses are high. However, once a large population inversion has been created in the active medium, the intensity of emission increases to the point where the dye is bleached, and stimulated emission can result in emission of a pulse of light. This last method of Q-switching is known as *passive*, since the switching time is not determined by external constraints.

Q-switching generally produces pulses which have a duration of the order of 10^{-9}–10^{-8} s. The pulse repetition rate is determined by the time taken to re-establish population inversion, and thus depends amongst other factors on the pumping rate, but the interval between pulses is typically a few seconds. In the case of a laser such as the ruby laser which is inherently pulsed, it is worth noting that Q-switching does somewhat reduce the energy content of each pulse. However, this disadvantage is more than outweighed by the enormous reduction in pulse duration, which thus results in a sizeable increase in the peak irradiance. If we ignore the energy loss, a rough calculation shows that for a laser producing 1 ms pulses of energy 10 J, in other words, a mean output power per pulse of 10 kW, then Q-switched operation which results in compression of the pulse to 10 ns increases the mean pulse power to 1 GW. Note that such calculations are necessarily crude because the intensity of emission

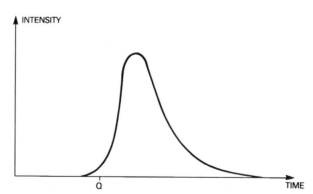

Fig. 3.10 The temporal profile of a Q-swiched laser pulse; the point Q on the time-axis denotes the onset of Q-switching

does not remain constant throughout the pulse duration; the temporal profile of a typical Q-switched laser pulse is illustrated in Fig. 3.10. Hence the *peak* intensity may be far greater than the computed mean.

3.3.3
Mode-Locking

The third widely used method of producing laser pulses is known as mode-locking. This method produces pulses of much shorter duration than either cavity dumping or Q-switching, typically measured on the picosecond (10^{-12} s) scale and referred to as *ultrashort*. The technique has very important applications in chemistry, for example in the study of ultrafast reactions (see Sect. 5.3.4), and it is therefore worth examining in somewhat more detail. The mechanism of mode-locking is as follows.

We have previously noted that laser emission usually occurs over a small range of wavelengths which fall within the emission linewidth of the active medium (see Fig. 1.14, p. 22), and satisfy the standing wave condition

$$m\lambda/2 = L, \tag{3.4}$$

where m is an integer. Hence, the emission frequencies are given by

$$\nu = mc'/2L, \tag{3.5}$$

and they are separated by the amount

$$\Delta\nu = c/2L, \tag{3.6}$$

(the free spectral range) which is typically around 10^8 Hz. In the case of solid-state or dye lasers, where the fluorescence linewidth is quite broad, the number of longitudinal modes N present within the laser cavity may thus be as many as 10^3–10^4 if frequency-selective elements are removed.

In such a situation, however, there is not generally any correlation in phase between the various modes, so that the effect of interference usually results in a continual and essentially random fluctuation in the intensity of light within the cavity. The technique of *mode-locking* consists of creating the phase relationship which results in completely constructive interference between all the modes at just one point, with destructive interference everywhere else. Consequently, a pulse of light is obtained which travels back and forth between the end-mirrors, giving a pulse of output each time it is incident upon the semi-transparent output mirror.

It can be shown that the pulse duration, expressed as the width of the pulse at half power, is given by

$$\Delta t = 4\pi L/(2N+1)c', \tag{3.7}$$

and the interval between successive pulses is obviously the round-trip time given by Eq. (1.21), i. e $\tau = 2L/c'$. Hence mode-locked pulses are typically

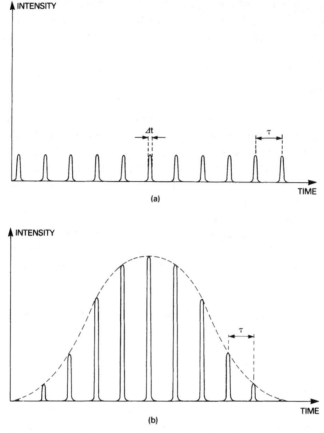

Fig. 3.11a and b The temporal profile of mode-locked emission from (a) a cw laser, and (b) a pulsed laser. In the second case, very many more mode-locked pulses than illustrated would normally be found within the overall pulse envelope

separated by an interval of around 10 ns, and have a duration of 1–10 ps; obviously the greater the number of modes the shorter the pulse length, in accordance with the Fourier relationship between pulse widths in frequency and time. The temporal profile of the emission from mode-locked cw and pulsed lasers is thus as shown in Fig. 3.11, and in each case consists of a pulse train. The instantaneous intensity of a mode-locked laser pulse can be readily computed on the essentially correct assumption that the overall energy emitted within the duration of one laser round-trip is the same before and after mode-locking. Thus with a pulse-length reduction from 10 ns to 1 ps, the pulse power can be increased by a factor of approximately 10^4, so increasing the power of gigawatt pulses up to the region of 10^{13} W.

As with Q-switching, there are several ways to accomplish mode-locking, involving both active and passive methods. The active methods generally involve modulating the gain of the laser cavity using an electro-optic or acousto-optic switch driven at the frequency c/2L. In much the same way as Q-switching, such elements placed inside the cavity can effectively act as shutters and for mode-locked operation become transparent only for brief intervals separated by an interval of 2L/c. Hence only light that is travelling back and forth in phase with the modulation of the shutter can experience amplification, which naturally results in mode-locking.

Passive mode-locking is generally accomplished by placing a saturable absorber cell within the laser cavity as in Fig. 3.9 c (though in the exceptional case of the titanium:sapphire laser, automatic mode-locking can be observed without the inclusion of any such element – banging on the table can often do the trick!) One of the most widely used saturable absorbers is 3,3′-diethyloxadicarbocyanine iodide, usually known as DODCI. Figure 3.12 illustrates how use of a saturable absorber results in mode-locked operation. At time t, the graph shows the initially random fluctuation of intensity along the length of the laser cavity, resulting from interference between different longitudinal modes. After a few round-trips back and forth, at time t′, the intensity pattern is as shown in the second graph. The effect of each passage through the saturable absorber cell is to reduce the intensity of the high amplitude fluctuations to a far lesser extent than the small fluctuations. The loss of intensity by the highest amplitude fluctuations is more than compensated by the amplification they experience on passage through the gain medium; however, the low amplitude fluctuations experience a net loss. Hence, in the course of each round-trip, the high intensity fluctuations essentially grow at the expense of the small fluctuations. Thus after several more round-trips, at time t″, the intensity pattern develops to the form shown in the third graph. Here, only one intense pulse remains, having grown from the initial fluctuation of the largest amplitude; all the other fluctuations have died away. Again, the result is mode-locked operation.

One other mode-locking technique principally employed with tunable dye lasers is *synchronous pumping*. Here a mode-locked primary laser pumps a dye laser whose optical cavity length is adjusted to equal that of the pump, so that the round-trip time of the dye laser matches the interval between pump pulses. In this way, frequency-converted pulses propagating within the dye laser cavity arrive at the the dye jet at the same rate as the pump pulses. The period over which amplification is effective in the dye laser is thus very short, and once again picosecond pulses emerge. In practice, the dye laser cavity length needs to be kept constant to within a few microns, so that special mounting and temperature control are necessary.

Several further points concerning mode-locked pulses are worth noting. First, it should be noted that although the method produces a train of pulses, it is possible to select out single pulses by using a Pockels cell and polariser

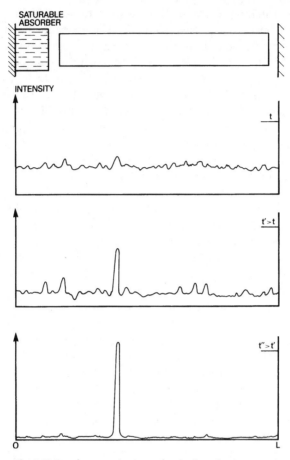

Fig. 3.12 Development in time of a single pulse in a passively mode-locked laser. The times t, t' and t" are separated by an integer number of round-trips in the cavity

combination as discussed earlier (Sect. 3.3.2). Here, however, both are placed outside the laser cavity, and the Pockels cell is activated by an electrical spark of shorter duration than the pulse separation τ; hence, only a single pulse from the mode-locked train can pass through. It is also quite common to employ mode-locking in conjunction with radio-frequency cavity dumping, so as to obtain variable repetition-rate trains of picosecond pulses. Because cavity dumping releases large amounts of cavity energy, the power levels of cavity-dumped mode-locked pulses are appreciably higher than those obtained by mode-locking alone, often by a factor of about 30.

There are various means by which pulse lengths can be reduced, as for example in *colliding-pulse* ring dye lasers. By use of other pulse compression techniques in conjunction with mode-locking, commercially available lasers

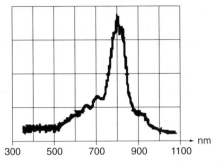

Fig. 3.13 Typical spectrum of the supercontinuum generated by a titanium: sapphire laser, drawn on a logarithmic vertical scale. (Redrawn from an original courtesy of Coherent Laser Group)

can now provide pulses of a few picoseconds duration in the infra-red, and ultrashort pulses measured on the femtosecond (10^{-15} s) timescale in the uv/visible range. The duration of pulses less than 10^{-4} s long can be meaningfully described in terms of a few optical cycles (visible frequencies being in the 10^{15} Hz range). The time-frequency uncertainty principle $\Delta\nu\Delta t \geq 1/2\pi$ demonstrates that a very broad span of frequencies is present in such pulses, so that pulse compression has to be at the expense of monochromaticity.

This brings us to another key aspect of ultrashort laser pulses. In 1970 Alfano and Shapiro[4] discovered that on passing picosecond laser pulses through certain media, a broad continuum of light was generated, typically covering a range of about $10\,000$ cm^{-1} about the laser frequency; moreover, the duration of the continuum pulse was of the order of 100 fs. The effect turned out to be surprisingly easy to produce; even focussing a mode-locked laser into a beaker of water will do the trick. The facility to produce such a broad continuum of light with such short duration has a number of distinctive uses, as for example in the flash photolysis study of ultrafast reactions (see Sect. 5.3.4). The light produced by this method, which results from a mechanism known as self-phase modulation, is referred to as an *ultrafast supercontinuum laser source*, or *picosecond continuum* for short – a typical example of its spectral profile is shown in Fig. 3.13.

Quite apart from the chemical applications of ultrashort laser pulses, which mostly involve time-resolved spectroscopy in the uv/visible range, there are also some very significant applications in physics which it is worth drawing attention to before leaving the subject. One is the possibility of inducing nuclear fusion reactions by making use of the enormously high intensities available. If pellets of lithium deuteride, for example, are encapsulated in thin spherical glass shells and irradiated by pulses of say 10^{21} W m^{-2} intensity, instanta-

---

4) Alfano R R, Shapiro SL (1970) Phys. Rev. Letts 24: 584

neous pressures of about 10^{12} atm can be created, which are sufficient to compress the nuclei together and induce nuclear fusion. A great deal of effort is being directed towards the development of this method as a means of producing clean, cheap energy in a controllable way. Other uses have been found in ultrafast photography and in the field of optical communications, where by using a simple pulse code modulation (PCM) technique to switch on or off each pulse in a mode-locked train, at least a million simultaneous telephone conversations can in principle be transmitted by a single laser.

3.4
Detectors

A wide range of optical detection equipment is employed in laser applications. A number of these detectors are specifically employed in characterisation of the input beam, mostly with regard to measurement of the wavelength, bandwidth, power, beam quality and beam profile factors discussed in Sect. 1.6. Here, however, we shall focus on detectors directly involved in signal acquisition. In virtually all laser measurements of a physical or chemical property, an electrical signal is produced by the response of some photodetector to light incident upon it. Several types of detector are used for this purpose; the exact choice for any particular application depends on the region of the electromagnetic spectrum and the power level involved.

First we consider *thermal* photodetectors. These operate by electrical detection of the variation in a physical property caused by heating. For example, electrical resistivity, gas pressure and thermoelectric emission can all be monitored for this purpose. Thermal detectors suffer limitations due to the nature of the primary mechanism whereby the energy of absorbed light is converted into thermal energy. Consequently, they often have comparatively slow response times; also the wavelength range over which they operate may be restricted by the absorption properties of the material used for the window. The most common type of thermal photodetector is the *pyroelectric detector* whose response is based on measurement of thermally-induced changes in electrical properties. Devices of this kind cover the entire wavelength range 100 nm–1 mm and respond to pulse frequencies of up to 10^{10} Hz.

A second class of laser detectors generally termed *photoemissive* operates on the photoelectric effect. All such detectors have a long-wavelength cut-off, in other words, a wavelength beyond which the photon energies are insufficient to release electrons, and therefore induce no response. The simplest device of this kind is a *vacuum photodiode*, in which electrons are released as light impinges upon a photosensitive cathode inside a vacuum tube, producing an electrical signal as they are received by an anode plate. The precise wavelength range over which such a device is effective depends on the nature of the photocathode material, and the wavelength for maximum sensitivity ranges from 240 nm for non-stoichiometric Rb/Te to 1500 nm for germanium. Since solid-state electronics is involved in the detection process, the response times of semiconductor detectors are very fast and are generally limited only by the associated circuitry.

Photomultiplier tubes are based on the same principle, but incorporate electronics which multiply the number of electrons received at the anode by a factor of between 10^6 and 10^8; thus in certain designs the absorption of even a single photon generates sufficient current (up to 100 mA) to be easily detected. This sensitivity is obviously highly significant for applications to the measurement of weak fluorescence or light scattering. Photomultiplier devices are most sensitive for the uv/visible region 185–650 nm, but generally have the drawback of a comparatively slow response time on the nanosecond scale. Nonetheless, the fastest devices can respond to pulse repetition frequencies of up to 10^{11} Hz. One other factor which limits the sensitivity of photoemissive devices in general is the so-called 'dark current', the residual current due to emission in the absence of irradiation. This is a strongly temperature-dependent phenomenon, which can be ameliorated by cooling the photomultiplier housing.

In *photoconductive* semiconductor sensors, the underlying principle is the excitation of valence electrons into a *conduction* band by the absorption of light. Once again, such detectors have a long-wavelength limit beyond which they are unusable. Here a very wide range of materials is employed, many of which are responsive over infra-red regions only. However, trialkali antimonides such as Cs_3Sb and Na_2KSb provide a good coverage of much of the visible spectrum, and some silicon devices are sensitive down to 180 nm. Most sensors have a non-stoichiometric composition tailored to provide response over a particular range. It is often necessary, when detecting infra-red radiation, to cool such detectors in order to eliminate the thermal excitation which would otherwise take place even in the absence of radiation. Occasionally, dry ice is sufficient, but frequently liquid nitrogen or even liquid helium may be required. Some photoconductive sensors can respond to pulse repetition rates of up to 10^{10} Hz, but somewhat lower frequency limits are more common; once again, response times are primarily determined by the signal processing circuitry. The spectral response curves for three common semiconductor detectors are shown in Fig. 3.14. Many detectors are now available linked to amplification and signal-processing electronics and are known as power or energy meters.

Finally, mention should be made of the comparatively new *array detectors* that have had a major impact on low light level spectroscopies, particularly Raman scattering (see Sect. 4.4). Photodetector arrays enable signals associated with a range of wavelengths to be simultaneously collected, essentially removing the drawback of slow data acquisition associated with slit-scanning spectrometers. The best performance array detectors are based on charge-coupled device (CCD) technology, typically offering wavelength coverage right across the visible spectrum and well into the near infra-red. Most of these

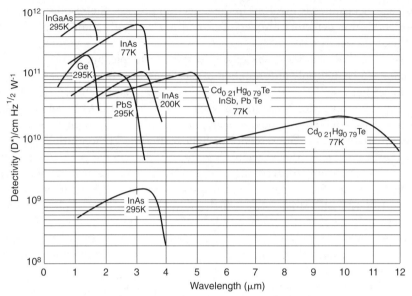

Fig. 3.14 Spectral responses of common semiconductor detectors, on a logarithmic scale. (Reproduced by kind permission of Optilas Ltd)

devices comprise a rectangular array of 'gates' laid over a suitably doped silicon substrate, the signals from each such pixel being rapidly processed to generate the spectral data. With liquid nitrogen cooling to reduce the levels of electronic noise, such devices are capable of producing excellent signal-to-noise ratios. Operated at the fastest rates of electronic scanning, as appropriate for higher levels of light intensity, CCDs are also usefully employed in many imaging applications.

3.5
Pulse Detection Systems

In chemistry, laser instrumentation is usually based on detection of the response of a material to laser irradiation. The response is monitored with any suitable detector such as a photodiode or photomultiplier tube which produces an electronic signal, and the signal is then amplified and fed to a display device. For example, in many spectroscopic applications a pen recorder is used, so that the response can be plotted against wavelength. It is outside the scope of this book to give fuller details of the electronics generally involved in detection instrumentation. However, there are certain additional features which have to be employed where pulsed irradiation is concerned, and it is worthwhile outlining these before proceeding to a discussion of specific applications.

3.5.1
Lock-in Amplifiers

One of the methods of increasing sensitivity in certain applications of *continuous-wave* lasers is to use a beam chopper and a lock-in amplifier (also known as a *phase-sensitive detector*) as shown in Fig. 3.15. Rotation of the chopper imposes a square-wave modulation on the intensity of light reaching the sample; the frequency of modulation is generally in the kilohertz region, but can lie anywhere between 5 Hz and 20 kHz, depending on the nature of the application. The lock-in amplifier is fed two signals; one is a control signal from a detector connected to the chopper, and the other is a response signal from the sample monitoring device. The job of the lock-in amplifier is then to selectively amplify only components of the response signal that oscillate at the same frequency as the control signal.

The result is a very effective discrimination against noise, and hence a large increase in the signal-to-noise ratio. Use of a lock-in amplifier with a dynamic range of 10^3 enables weak signals to be detected even when the level of noise is one thousand times greater. In fact, by electronically pre-filtering the input to discriminate against noise frequencies away from the frequency of interest, the dynamic range can be improved to as much as 10^7. Hence, uses arise, for example, in the detection of very weak laser-induced fluorescence.

The lock-in method is also commonly used when an inherently weak *effect* is monitored, as for example in optogalvanic spectroscopy (Sect. 4.2.5). One other more general kind of spectroscopic application lies in the *ratiometric recording* of absorption spectra. Here, the laser beam is split into two paths, one of which is modulated with a chopper and passes through the sample and the other of which is separately modulated and used as a reference. Both beams are then detected and produce signals which can be ratioed in conjunction with a lock-in amplifier. In this way the sample spectrum is automatically corrected for the variation of source output and detector efficiency with wavelength.

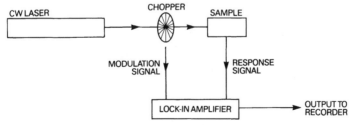

Fig. 3.15 Improvement of signal sensitivity by use of a chopper and lock-in amplifier

3.5.2
Boxcar Integrators

A boxcar integrator or *gated integrator* is a device widely used in conjunction with a pulsed laser producing a regular train of pulses, as for example in the case of a mode-locked source. The response signal from the sample is then also a periodic function; a typical response is represented in Fig. 3.16. There are two different modes of operation of the boxcar. First. as shown in (a), the device can be used as an electronic gate which responds only to signals received during a period Δt comparable to the duration of the pulse response. The signal received during this period is integrated and averaged over many pulses to produce the boxcar output. This method is the simplest to use and is appropriate when the time-dependence of the response is not of interest. It results in a better signal-to-noise ratio than direct signal integration since the noise usually present during the intervals between pulses is completely eliminated.

The second boxcar method, represented by Fig. 3.16b, involves sampling the response signal at a rate which is slightly different from the pulse repetition rate, with integration over comparatively short time intervals Δt. Thus different portions of the response curve are sampled from each successive pulse, and after a sufficiently large number of pulses the boxcar produces an averaged output which registers the signal level at various points through a typical pulse response. This approach is particularly useful in the determination of fluorescence decay lifetimes.

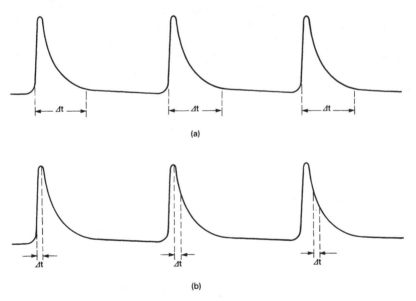

Fig. 3.16a and b Use of a boxcar integrator: (a) to average the response from a series of laser pulses, and (b) to sample the response at various points during the decay time of a single pulse

3.5.3
Single-Pulse Systems

In experiments involving the study of laser-induced physical or chemical change in a sample, it is often necessary to monitor the effect produced by one single laser pulse. One method is to use a *transient recorder*, an electronic device which retains a temporal representation of the response signal. This is accomplished by digitally storing in a memory the signal level at a large number of closely spaced but equidistant time intervals. Time resolutions down to 100 ps are obtainable with this type of device. However, for the measurement of ultrafast processes such as the primary events in photosynthesis, resolution needs to be much better than this. The device which is capable of the best time resolution is known as the *streak camera.*

Early streak cameras operated by focussing light onto a photographic film shot at very high speeds through a camera. The lateral displacement along the film thus represented a measure of the time elapsed, and by using a microdensitometer on the recorded image the intensity could be plotted as a function of time. As the duration of the shortest laser pulses has been reduced further and further, new methods have been required for their measurement. Today, the name 'streak camera' is usually applied to an electronic instrument operating on the same basic principle, but capable of resolutions down into the subpicosecond range. As illustrated in Fig. 3.17, the device is a modified cathode-ray tube, which by application of a sweep voltage to the deflection plates allows pulse duration and intensity to be calculated from the spatial extent of the image and its density on the screen.

Whilst streak camera methods are appropriate for the study of ultrashort laser pulses per se, or in some cases for laser measurements of fluorescence kinetics, other techniques can be adopted for the study of ultrafast processes directly connected with the *absorption* of ultrashort pulses. A common setup is illustrated in Fig. 3.18 and involves use of a beam-splitter to create two pulses from one original ultrashort pulse; one of these is used to excite the sample and the other is used as a probe of the excitation. A variable delay is introduced between the two separated pulses using a stepping motor allowing the

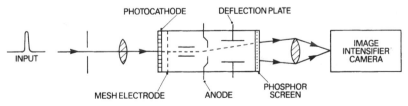

Fig. 3.17 The streak camera. Electrons released as the laser pulse strikes the photocathode are swept out electrically into an elongated image on the screen, whose length depends on the pulse duration

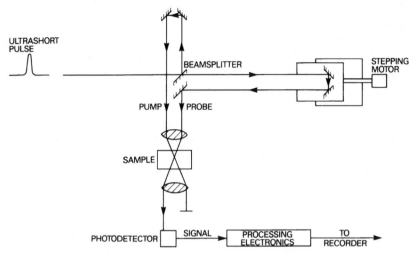

Fig. 3.18 Instrumentation for kinetics measurements over picosecond and sub-picosecond time intervals

difference in optical path length to be changed in small increments. Typically, spatial displacements as small as 0.1 μm can be introduced in this way, the corresponding 0.2 μm change in optical path length corresponding to a 0.7 fs delay. So, the distances involved in ultrafast measurements are in just the right range for very precise and yet very simple experimental control. The absorption from the probe pulse can be monitored at various ultrashort time intervals, thus providing information on the kinetics of the processes induced by the initial excitation. The delayed pulse can also be used to generate an ultrafast supercontinuum, as discussed earlier (Sect. 3.3.3), making it possible to probe the response of the sample over a wide range of wavelengths.

Having discussed some of the instrumentation for modification and detection of laser radiation, we now move on to consider the more specific instrumental considerations for particular applications. The techniques discussed in the following sections of this chapter are not explicitly spectroscopic in the sense that a spectrum is necessarily recorded, although spectroscopic principles may be involved. Genuinely spectroscopic applications are discussed in Chap. 4.

3.6
Light Scattering Instrumentation

Many chemical applications of lasers involve irradiation of a sample and detection of its absorption or subsequent fluorescence. Nonetheless, a great deal can be learned from analysis of the light which a sample simply scatters. Most of the light which is scattered has essentially the same frequency as the laser light itself; this is known as *Rayleigh scattering*, and is itself a weak effect, seldom accounting for more than one photon in 10^4. The very small frequency shifts which are associated with Rayleigh scattering result from *Doppler shifting* due to the motion of sample molecules.

A much smaller amount of light is nevertheless inelastically scattered with somewhat larger shifts in frequency, and the monochromaticity of the laser radiation facilitates its detection. Inelastic light scattering comes under two main headings: *Brillouin scattering*, in which energy is transferred to or from acoustic vibrations in bulk materials, and *Raman scattering*, in which it is generally transitions in individual molecules which are responsible for the loss or gain of photon energy. In this section we examine some of the analytical applications of these light scattering processes; we return to a much more detailed look at the principles of Raman *spectroscopy* in the next chapter.

3.6.1
Nephelometry

The elastic scattering of light without change in frequency occurs both from free molecules and from suspensions of larger aggregate particles. The term 'Rayleigh scattering' is more commonly applied to the former case or to particles of diameter not exceeding roughly a tenth of the wavelength of light involved. Such scattering, whilst not isotropic, nonetheless occurs over all directions, and the angular distribution of scattering intensity provides a crude 'fingerprint' of the medium. This method has, for example, been suggested as a simple but objective means of characterising wines. The technique does not, however, provide much useful chemical information that cannot be more easily obtained by other means.

The more intense scattering of light observed from larger particles has different characteristics and is especially effective close to the forward-scattering direction. When the size of each particle is sufficiently large that it is no longer transparent, the nature of the effect changes once more as surface reflection takes over. *Nephelometry*, which actually means the measurement of cloudiness, is a quantitative measure of the scattering of light from aggregate particles such as those in colloidal suspension. The instrumentation simply involves placement of a photodetector beside the sample, but out of the direct pathway of the laser radiation. The technique can be especially useful in conjunction with liquid chromatography, as discussed in Sect. 3.8.

3.6.2
Photon Correlation Measurements

Whilst measurement of the gross intensity of Rayleigh scattering provides a certain amount of data on a sample, much more detailed information on solutions and liquid suspensions can be obtained by monitoring the intensity *fluc-*

tuations in the scattering from a cw laser beam. Assuming that the laser has good amplitude stability, such fluctuations primarily result from the essentially random interference of light scattered by different molecules or molecular aggregates at the focal point in the sample. Typically the scattered light is detected by a photomultiplier and the signal is processed in a *correlator*, which determines the timescale over which fluctuations take place. The results can ultimately be used to characterise various dynamic properties of the sample.

The quantity actually measured by a correlator fed two time-varying input signals I(t) and J(t) is the *correlation function* $G(\tau)$, defined by the formula:

$$G(\tau) = \lim_{\tau \to \infty} (1/2T) \int_{-T}^{T} I(t)J(t + \tau)dt. \tag{3.8}$$

Whilst some *heterodyne* measurements do indeed feed two different signals to the correlator, one derived directly from the laser and the other from the scattered light, *homodyne* measurements are more common, in which I and J are the same signal and G represents the *autocorrelation function* of the scattered light. The technique by which this autocorrelation function is derived is variously known as *quasi-elastic light scattering, dynamic light scattering, intensity fluctuation spectroscopy,* and *photon correlation spectroscopy.* The term 'spectroscopy' is basically a misnomer, since it is not the frequency distribution of the scattered light that is directly measured; nonetheless, the spectral distribution of the scattered light can, if desired, be obtained from the correlation function by Fourier transform. This method is particularly useful for measurement of very small frequency shifts of less than 1 MHz. The laser used for photon correlation measurements must operate in the TEM_{00} mode, have good amplitude stability, and emit a suitable wavelength that is not absorbed to any significant extent by the sample. For low levels of scattering, the principal 488.0 and 514.5 nm lines from an argon laser are frequently employed; for stronger scattering, 632.8 nm radiation from a He-Ne laser is more commonly used. Typical correlator apparatus can measure intensity fluctuations with a time-resolution of down to 50 ns.

For solution studies, the only constraint on the sample itself, apart from transparency at the laser wavelength, is a mismatch between the refractive indices of the solute and solvent. It is worth noting that where the scattering intensity is especially weak, data collection times of several hours may be necessary. In such cases, it is of course imperative that the liquid sample is scrupulously clean, since dust, bacteria, and any other foreign matter can produce strong spurious fluctuations in the scattered light. Various kinds of data are available by use of this relatively cheap technique. For example, the rotational and translational diffusion coefficients of macromolecules in solution can be ascertained, which in turn provide valuable information on molecular shape and dimensions. Such information is especially useful in the study of polymers, since it can be related to their degree of polymerisation and mean molecular weight. Similar measurements on micellar systems (colloids) provide data on micellar shape, aggregation number, and micellar interactions. This facilitates the characterisation of biological micelles and synthetic aqueous detergents, for example. There is also a wide range of medical and physiological applications, ranging from immunoassay to determination of the motility of bacteria and spermatozoa.

Finally, it is worth noting that autocorrelation measurements based on second harmonic generation are widely used to measure pulse lengths of mode-locked lasers. Here an interferometer set-up is used, similar to that depicted in Fig. 3.18, with a suitable non-linear crystal replacing the sample. The measured harmonic intensity depends on the pulse autocorrelation function, thus enabling the pulse length to be deduced if its temporal profile is known.

3.6.3
Brillouin Scattering

As mentioned earlier, Brillouin scattering results in a shift in scattering frequency due to coupling between the radiation and acoustic modes of the sample. The effect is strongest in solids and liquids, but can be observed weakly in gases. Since acoustic frequencies are very low compared to optical frequencies, the resultant shifts are correspondingly small, typically being of the order of 1 cm^{-1} in wavenumber units. Brillouin scattering cannot therefore be resolved from Rayleigh scattering using conventional filters or monochromators; instead, a *Fabry-Perot interferometer* has to be employed. This generally consists of two parallel glass plates each coated with a thin film of silver, one of which is stationary and the other of which can be moved towards or away from it on a carriage. Light travelling back and forth between the two plates undergoes multiple reflections rather like those within a laser cavity, and for any particular separation only a single wavelength can propagate through the system without suffering destructive interference.

Typical apparatus for the measurement of Brillouin scattering is illustrated in Fig. 3.19a. The source is generally a highly monochromatic ion laser, in which the effects of Doppler broadening are obviated by an intracavity etalon restricting emission to a single mode. Light scattered from the sample is resolved in the interferometer and gives rise to a spectrum such as that shown in Fig. 3.19b. Various types of information can be deduced from the results. In liquids, Brillouin scattering provides one of the few methods of obtaining information on the low-energy interactions which determine local structure; energy transport mechanisms in gases can also be studied in the same way. In the case of crystalline solids, each Brillouin feature represents coupling with one of the acoustic-frequency lattice vibrations known as *phonons* and yields information on structure and binding forces; such measurements are important in the semiconductor device industry, for example. Other quite different kinds of solid usefully investigated by this method include biological macromolecules

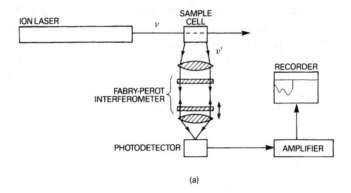

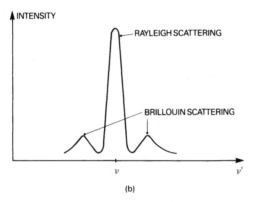

Fig. 3.19a and b Measurement of Brillouin scattering: (**a**) typical instrumentation using a Fabry-Perot interferometer; (**b**) a typical spectrum

and, in particular, fibrous materials such as collagen. Here, data on the elastic properties can be interpreted in terms of chemical properties such as the extent of hydration.

3.6.4
Doppler Velocimetry

In Chap. 1, we examined one manifestation of the Doppler effect, in the shift in the frequency of light emitted by molecules due to their Brownian motion. The same principle applies to the scattering of light in any direction other than forward scattering. However, bulk flow of the sample at any direction other than at right-angles to the beam also produces a Doppler shift which can be measured using the setup shown in Fig. 3.20. Here, use is made of the coherence of the laser light to produce interference between the beam and radiation back-scattered from the sample. This interference produces a beat frequency which can be electronically measured and directly related to the velocity. Velo-

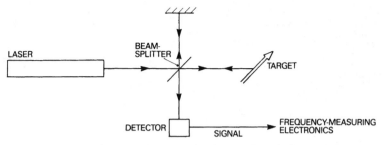

Fig. 3.20 Instrumentation for Doppler velocimetry

cities up to 40 ms^{-1} can be measured in the laboratory with commercially available instruments, although higher speeds can be measured at a distance; the technique is, for example, applied in the speed checking equipment used by the police. Even wind velocity can be measured by this technique (Doppler anemometry). Chemical uses of the principle usually involve fluid flow measurement at velocities down to 100 μm s^{-1}; there are also important biological applications for the non-invasive measurement of blood flow.

3.6.5
Lidar

A very different type of laser instrumentation is found in 'lidar', an acronym for *light detection and ranging*, where once again light scattering is the principle involved. This is a method primarily used for the remote sensing of trace chemical species in the atmosphere and is an important tool in pollution control. The essential details of the apparatus are shown in Fig. 3.21a. A pulsed laser emits short pulses of light directed into the atmosphere, and a telescopic detection system usually placed alongside the laser picks up any light which is back-scattered. The time interval between the emission of each pulse and the subsequent detection of scattered light is a direct measure of the distance of the substance responsible for the scattering. Hence, a plot of the detected intensity against time provides information on the variation of concentration with distance, as illustrated in Fig. 3.21b.

Various lasers are used for this type of application, usually Q-switched and emitting pulses of the order of 10 ns duration. Since the only other limitation on ranging precision is the response time of the detection electronics, also typically measured in nanoseconds, the distance of the scattering species can easily be measured to an accuracy of a few metres. Where the species of interest is in the form of aerosol particles, use can be made of the fact that such particles scatter light one or two orders of magnitude more effectively than gases. Here, then, a large response from the detector signifies the presence of the pollutant. For this purpose, ruby or Nd:YAG lasers are most often used. The

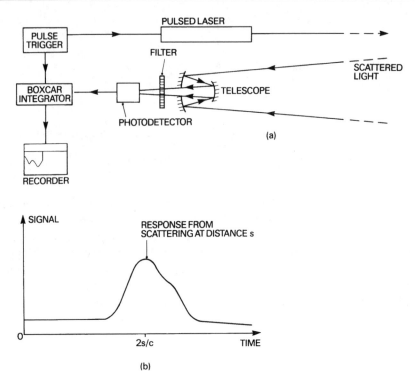

Fig. 3.21a Essential features of a lidar system, and (**b**) the response measured from the time of laser pulse emission

high degree of collimation of the laser beam means that scattering can be detected over a large range of distances, up to 100 km in the case of clouds.

There are several chemically more specific means of using the principles of lidar for atmospheric analysis. One possibility is the measurement of laser-induced fluorescence, which for good range resolution necessitates population of very short-lived excited states in the species of interest. This technique is therefore only practicable with uv/visible lasers which populate electronic states of nanosecond lifetime. A far more attractive method is to infer the presence of a pollutant from the reduction in scattering intensity when the laser wavelength coincides with one of the compound's absorption bands. Usually, the scattering is monitored at two different laser wavelengths, one resonant with an absorption band of the pollutant and the other not; this technique is known as *differential absorption lidar*. The differential absorption of the resonant wavelength over a particular range of distance thus demonstrates the presence of the pollutant in a particular vicinity, and the concentration can be deduced from the relative reduction in scattering intensity. Carbon dioxide lasers are widely employed for their power and range, though tunable diode lasers sometimes offer the possibility of more closely matching a suitable absorption wavelength of a particular pollutant. In certain cases it proves more expedient to monitor absorption in the ultraviolet. For example, the pair of wavelengths 286 and 300 nm, produced by doubling the output of a dye laser pumped by a frequency-doubled Nd:YAG laser, provide a basis for airborne measurements of stratospheric ozone concentrations and as such have been used in studies of the Antarctic ozone hole. Using the 300-nm absorption as reference, the localised reduction in 286 nm absorption not only provides the basis for measuring the diminished ozone concentration but also directly points to the disconcerting increase in the atmospheric transmission of ultraviolet sunlight. Concentrations as low as 0.1 ppm are measurable by this method.

A more sophisticated version of lidar instrumentation for gas detection is based on the Raman effect, described in detail in Chap. 4, and relies on the characterisation of pollutants by their molecular vibrations. The setup requires use of a uv/visible laser source and placement of a monochromator between the telescope and photodetector. In this way Stokes-shifted Raman wavelengths can be detected, thereby enabling the chemical composition of the pollutant to be identified unambiguously. Indeed, the temperature of the gas may itself be calculated from the intensities of the rotational Raman lines. All such lidar methods have the enormous advantage of enabling range-sensitive chemical analysis of the atmosphere to be performed from ground level. One of the prime areas of application is in the monitoring of stack gases from industrial chimneys. The remote lidar of the atmosphere above volcanoes has also been used as a predictor of volcanic activity, based on the known correlation of such activity with the sulphur dioxide content of the volcanic gases.

3.7
Polarimetry

When plane polarised light passes through a chiral substance, it experiences a rotation of its plane of polarisation known as *optical rotation*. In the case of isotropic liquids or solutions, the requirement for optical rotation to occur is that the solvent or solute molecules are in themselves chiral. In other words, they must possess no centre of symmetry nor any reflection- or rotation-reflection symmetry. Such molecules are known as *optically active* and include most large polyatomic compounds. The quantitative measurement of the extent of polarisation rotation is known as *polarimetry* and can be used to provide information on molecular optical activity, solution concentration, and path length. The parameter in terms of which most optical rotation results are expressed is the specific optical rotatory power, or *specific rotation* $[\alpha]$, defined by

$$[\alpha] = \theta/cl, \tag{3.9}$$

where θ is the angle of rotation, c the concentration and 1 the path length through the solution. The *molecular rotation* is obtained on multiplying by

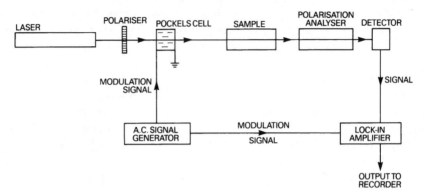

Fig. 3.22 Instrumentation for polarimetry

the molecular weight. Additional information characterising the particular substance involved can be obtained by measuring the dependence of optical rotation on the wavelength of light used for its detection; this technique is known as *optical rotatory dispersion.*

Lasers are ideally suited to polarimetric analysis. A typical setup is illustrated in Fig. 3.22 and involves polarising optics and pulsing techniques discussed earlier in this chapter. Also involved is a polarisation analyser, which is essentially a variable-orientation polariser. Since the minimum amount of light passes through this polariser when it is set at 90° to the plane of polarisation of the incident light, rotation of the analyser enables the orientation of the polarisation plane to be determined.

The instrumentation is easiest to understand if we first consider the setup with no sample present. The Pockels cell modulates the orientation of the polarisation plane, resulting in modulation of the intensity of light passing through the analyser and on to the photodetector. However, when the polarisation plane of the analyser lies exactly inbetween the two extreme polarisations emerging from the Pockels cell, the photodetector signal oscillates sinusoidally at exactly twice the modulation frequency. Now with the sample present, the same condition is only fulfilled if the analyser is rotated through some small angle; this measures the optical rotation of the sample. Rotations as small as 0.003 mrad can be routinely detected with this type of apparatus, although the best polarising optics can enable measurement of rotations almost two orders of magnitude smaller. The technique is quite amenable to application for very small volumes of solution and is thus particularly useful in conjunction with liquid chromatography.

Before leaving this subject, it is worth noting that *circular dichroism* measurements are also very amenable to laser instrumentation, using the same principles of electro-optic modulation of polarisation. In this case, it is the *differential absorption* of left- and right-handed circularly polarised light that is used to characterise chiral compounds. Results are often expressed in terms of Kuhn's *dissymmetry factor*, given by

$$g = (\varepsilon^{L} - \varepsilon^{R})/\bar{\varepsilon}, \tag{3.10}$$

where ε denotes the molar absorption coefficient (see Sect. 4.1), and $\bar{\varepsilon}$ is the mean of the results ε^{L} and ε^{R} for left- and right-circularly polarised light. The value of the dissymmetry factor for absorptions in the visible/uv region is typically of the order of 10^{-4}. Since the spectral lineshape associated with circular dichroism bands is often appreciably narrower than the corresponding optical rotatory dispersion curves, there is generally a better spectral resolution, and the circular dichroism technique is therefore often to be preferred.

3.8
Laser Detectors in Chromatography

For the detection of trace quantities of organic compounds in mixtures with other substances, one of the most powerful methods is chromatography. The general principle behind chromatography is the separation of different components of a mixture by their different rates of transport through various media. The most fully developed and sensitive type of chromatography, applicable to mixtures which are readily volatilised, is gas chromatography; the preferred method for relatively involatile liquid samples is high-performance liquid chromatography (HPLC). One of the difficulties with this type of chromatography has been finding a sufficiently sensitive detector; there is now every indication that lasers may provide the answer.

The methodology of HPLC is essentially similar to conventional liquid chromatography, in which a solution of the sample is carried by a solvent (eluant) under gravity down a column packed with a fine powder; different components of the solution travel through the column at different rates depending on their transport properties and affinity for the column material and, hence, have different and characteristic column retention times. The chromatogram is obtained by monitoring a suitable physical property of the solution eluted from the bottom of the column and plotting it as a function of time.

HPLC involves the same principles, but the column material is composed of much finer particles, typically 5–10 µm in diameter, and consequently the liquid has to be forced through under a high pressure, typically of the order of 100 atm. A preparation of hydrated silica particles is the most commonly used material for the separation, in which case a non-polar solvent is employed, and the most polar solutes tend to have the longest retention times. Alternatively, organically substituted silica may be used (the OH group most often replaced by straight-chain $OC_{18}H_{37}$) so that the solvent is more polar, and the tendency is for polar solutes to be retained the shortest times; this method is known as *reverse-phase* liquid chromatography.

Various laser techniques can be employed to monitor eluant from an HPLC column. Mostly, these entail spectroscopic principles which are described in detail in the next chapter. The narrow beamwidth and high intensity of laser light generally means that very small detection volumes can be employed; it is usually the flow design which is the limiting factor, but volumes as small as 10^{-8}l are in principle possible. By the same virtue, laser methods are also singularly appropriate for 'on-column' detection. Nonetheless, since very low concentrations are usually involved, the two most obvious methods of detection, namely absorption and fluorescence, are both applicable only where the substances of interest display appreciable absorption at the operating wavelength of the laser. Absorption features are generally quite broad in liquids, and the process of chromatography can itself provide a means for the separation and identification of the components in a mixture. One of the absorption methods which shows most promise for this purpose is thermal lensing spectroscopy (see Sect. 4.2.3). Where measurements are based on fluorescence, the sensitivity towards species which fluoresce only weakly can be markedly improved by 'tagging' with a strongly fluorescent label; this approach is often used for the detection of biomolecules. Despite the fact that fluorescence may occur over a wide range of frequencies, the monochromaticity of a laser detector is nevertheless important in these measurements since with suitable filters it facilitates discrimination against Rayleigh and Raman scattering. It has been demonstrated that picogram quantities can be detected with HPLC fluorescence instrumentation.

Measurements based on the Raman effect itself or the related method of CARS (see Sects. 4.4 and 4.5.4) also require highly intense and monochromatic sources, and lasers are the obvious choice. Here, the price of spectral resolution is a loss in sensitivity; these methods are therefore more useful for characterisation than for detection. Another option for the dection of optically active species is laser polarimetry. Finally, the principles of nephelometry can be employed. Since HPLC eluants are rarely turbid, this method is most successful if a precipitating agent is introduced before the detection stage.

3.9
Laser Microprobe Instrumentation

A number of laser applications in chemical analysis entail use of microprobe instrumentation. Such techniques are often used to probe the varying chemical composition of heterogeneous solid materials such as minerals or corroded metals. The instrumentation basically consists of an optical microscope, through which laser light is focussed onto the sample. As shown in Fig. 3.23, visual inspection through the microscope faciltates movement of the sample so as to probe its response at various sites. Microprobe methods thus have microsampling capability without the need for any sample preparation. They also

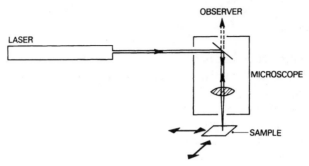

Fig. 3.23 Essential laser microprobe setup. The detection instrumentation depends on the spectroscopic principle employed

offer the advantages of real-time analysis and, in certain cases, the ability to obtain depth profiles of chemical concentration.

The principal disadvantage of a laser microprobe is its relatively poor spatial resolution, especially compared to electron microscopic methods. The resolution limit for a focussed laser is determined by the diffraction-limited beam width and is of the order of one wavelength. For a laser operating in the visible range, resolution can therefore seldom be less than about 0.5 μm (but see below). Various mechanisms are employed for analysis of the response of the sample to a microprobe laser. Trace element analysis is usually performed by using a powerful laser source to vapourise minute quantities of the sample. These are subsequently characterised by atomic fluorescence or mass analysis in a mass spectrometer. An alternative method which is entirely non-destructive involves measurement of the Raman scattering induced by the laser beam. These methods are discussed in more detail under the appropriate headings in the next chapter.

With suitably designed optical probes, it is in fact possible to image with a substantially better than diffraction-limited spatial resolution through a comparatively recent technique known as *near-field microscopy*. Another very different laser microprobe technique used for the chemical analysis of liquids and solutions, also based on optical fibre instrumentation, is *remote fibre fluorimetry*. Here, the basic principle is the passage of laser light along an optical fibre to a minute lens at its end known as an *optrode* immersed in the solution to be monitored. Fluorescence from ions or other species excited by the laser light then passes back through the fibre and can be detected and converted to an electrical signal in the usual way. The optrode can be designed to be ion-specific, by using a coating of a suitable chemical reagent to react with the ion and give a fluorescent product. Strongly fluorescent solution species can be monitored with a high degree of sensitivity using an optrode protected by a suitable semipermeable membrane. Detection limits in the parts-per-trillion range have been demonstrated using this fluorimetric method.

3.10
Laser Safety

Safety has in the last decade become more a matter for adherence to legislation than individual assessment. Since regulations differ from country to country, it is only appropriate to present a brief overview of laser safety. In fact, lasers have a remarkably good safety record, despite the fact that beams have in the past often been allowed to propagate freely through laboratory space. In such cases the room is often darkened to prevent stray light entering the detection system, and where visible wavelengths are concerned the high intensity and collimation of the laser beam usually make its path clearly visible through light scattering. There should be little scope for accidents in such situations.

Many of the hazards associated with laser operation are not directly connected with the beam itself. Indeed, the greatest danger often lies in the high-voltage power supply generally required for lasers and associated items of electro-optical equipment, and it appears that the majority of serious accidents so far incurred by laser users have involved electrocution. There are often other supplementary dangers, such as those connected with the cryogenic apparatus used for cooling high-powered cw sources; there is even a noise hazard in the case of a gas-dynamic laser. However, most such hazards can be dealt with by the adoption of obvious and well-established precautions. We shall therefore concentrate on the optical radiation dangers associated with the laser beam.

The greatest cause for concern is the possibility of the collimated laser beam either directly or by reflection entering the eye. Various types of damage can be caused, depending on the wavelength, intensity and exposure time. The precise mechanism for tissue damage in the infra-red and visible regions generally involves thermal effects, or even in some cases photoacoustic shock, whilst in the ultraviolet, damage is initiated by photochemical processes. With most laser radiation in the ultraviolet or far infra-red, there is an appreciable risk of accidental eye injury compounded by the invisibility of the light (although sometimes a visible wavelength is also passed through the system collinearly so as to make the beam pathway visible). Such radiation is not focussed onto the retina, but is absorbed by the cornea and lens, where it causes the damage. If only the outermost layer of the cornea is damaged, the normal process of tissue replacement means that despite the intense pain, complete healing may occur within a couple of days. Much more serious is the possibility of lens damage from the absorption of near ultraviolet or infra-red radiation. This may result in the formation of cataracts many years later.

Although the risk of injury is less likely with visible light, radiation anywhere in the visible/near infra-red region 400—1,400 nm is potentially more damaging since it is focussed onto the retina, and thereby suffers an increase in intensity. A human eye adapted to dark conditions has a focussing power of approximately 5×10^5. Since, for example with ruby laser radiation, energy densities of less than 1 kJ m^{-2} can cause retinal damage, then allowing a safety factor of 10 the maximum permissible exposure level at the cornea is of the order of 10^{-4} Jm^{-2}. A single pulse from a Q-switched ruby laser will thus cause severe tissue disruption at the retina; of course, this is the basis for some ophthalmic procedures. Diffusely reflected laser light is less of a problem; there, a safety limit of 100 J m^{-2} should be acceptable. Even moderately low-power cw lasers can cause severe retinal burns. The retina is not normally exposed to *continuous* intensities above 1 W m^{-2}; anything more intense results in aversion response. The Sun, for example, produces an image intensity on the retina of about 10^5 W m^{-2}. Retinal burns will occur before there is time for the natural blink response if the image intensity exceeds approximately 10^6 W m^{-2}. The resultant damage to the rod and cone cells produces an immediate and permanent degradation of both colour and night vision.

Skin on other parts of the body is most susceptible to injury over the so-called 'actinic' range 200–320 nm. However, unless there is prolonged exposure to beams exceeding several tens of kW m^{-2}, damage is unlikely to occur; usually the warmth response enables evasive action to be taken before this stage is reached. Skin hazards are therefore usually only associated with high-power beams. The maximum permissible exposure varies very much according to the wavelength, not least because of the variation in skin reflectivity. For example, 40% of the 1.06 μm radiation from a neodymium laser is reflected from an average skin surface, whilst the corresponding figure for 10.6 μm carbon dioxide radiation is only 4%. Thus an energy density of 4×10^5 J m^{-2} may be possible before a neodymium laser causes skin burns, whilst the carbon dioxide laser may cause damage at energy densities a factor of ten less. The exposure limit for radiation which is pulsed on the nanosecond or shorter time-scales may be still smaller, because there is less time for heat dissipation. For example, the maximum exposure figure for a Q-switched CO_2 laser is 2×10^3 J m^{-2}. Lasers are generally classified into one of four classes depending on the degree of hazard associated with the beam. Class I lasers, for which it is deemed that no safety precautions are necessary, include only low-power GaAs or similar semiconductor sources or other lasers enclosed within machine equipment. Class II lasers, such as low-power He-Ne lasers with output below 1 mW are dangerous only in the unlikely event of prolonged staring into the beam (*intrabeam viewing*). Class III lasers can be defined as those which can cause eye damage within the time taken for a blink response, which is about 0.2 s. However, such lasers are not hazardous to the skin under normal conditions. The principal consideration with this type of laser is that apparatus should be constructed in such a way that it is impossible to place the eye directly in line with the beam. Finally, we have class IV lasers which cause serious skin injury, and where even diffuse reflections can damage the eye. For this reason, protective goggles are traditionally worn by the laser operators. The heating effect associated with any such source may also be considered a fire hazard. All cw lasers with an output power above 0.5 W fall into this category and call for very comprehensive safety precautions.

3.11
Questions

1. Outline how you would convert the output of an argon ion laser operating at 488 nm (a) to higher wavelengths in the orange and (b) to a lower wavelength in the ultraviolet region.

2. Radiation from a forsterite laser operating at 1340 nm is passed through a Raman shifter based on barium nitrate, giving 30% output with a 1047 cm^{-1} Stokes shift. Calculate the wavelength of the shifted output.

3. Briefly describe how you would produce laser radiation at a wavelength of 560 nm using a Nd:YAG pump laser.

4. A krypton laser emitting at 647.1 nm is used to pump a dye laser based on the dye Rhodamine 700 and produces output at 801.3 nm. The pump beam and the dye laser beam are then focussed together onto a crystal of lithium niobate in which frequency addition takes place. What new wavelength is produced by this interaction?

5. The output of a laser operating at a wavelength λ is passed through a beamsplitter to produce two beams, (a) and (b). Beam (a) is passed through two successive frequency-doubling crystals giving an output (c). Beam (b) is used to pump a dye laser whose output (d) is shifted in wavelength by an amount $\Delta\lambda$. Finally beams (c) and (d) are combined by frequency-addition to produce an output of wavelength λ'. Derive a formula for λ' in terms of λ and $\Delta\lambda$.

6. At the molecular level, frequency-doubling and frequency-addition are both processes in which three photons are involved; two are absorbed and one is emitted in each conversion. The three interactions are nonetheless coupled in such a way that the medium in which either process takes place returns to its ground state with the completion of each conversion. By considering the implications of the Laporte selection rule for electric dipole transitions, show that these processes are necessarily forbidden in centrosymmetric materials.

7. Calculate the pulse duration $\Delta t = 4\pi L/(2N + 1)c'$, and the interval between pulses, $\tau = 2L/c'$, for a mode-locked laser of cavity length L=1 m sustaining N=2750 uniphase standing waves within its cavity. ($c' = 3.00 \times 10^8$ m s^{-1})

8. For a Nd:glass laser of optical cavity length 1.5 m (corrected for refractive index) the mode-locked output at 1.06 µm consists of 300 µJ pulses of 6.0 ps duration. Using Eq. (3.7), estimate the number of uniphase modes sustained in the cavity, and also the number of photons per pulse. Also estimate the instantaneous irradiance obtainable from each pulse if the beam is focussed down to a beam-width of 5 µm. ($c' = 3.00 \times 10^8$ m s^{-1}; $h = 6.63 \times 10^{-34}$ J s).

9. Using Eq. (1.24), calculate the beat frequency $(\nu - \nu')$ obtained from the laser Doppler velocimetric measurement of a liquid flowing coaxially with

the laser beam at 100 μm s^{-1}, if a helium-neon laser is employed (wavelength 633 nm).

10. The distance resolution of lidar is limited by the shortness of laser pulse duration and the uncertainty in the response time of the detector. By a graphical or any other method, calculate the range-resolution of a lidar system based on a Q-switched laser delivering 10-ns pulses, coupled with a detector having a response time uncertainty of 3.3 ns ($c' = 3.00 \times 10^8$ m s^{-1}).

CHAPTER 4

Chemical Spectroscopy with Lasers

> 'Why is the grass so cool, fresh, and green?
> The sky so deep, and blue?'
> Get to your Chemistry, You dullard, you!
> 'The Dunce', Walter de la Mare

Spectroscopy is the study of the wavelength- or frequency-dependence of any optical process in which a substance gains or loses energy through interaction with radiation. In the last chapter, we considered several strictly *non*-spectroscopic chemical techniques, mostly based on interactions with laser light at a *fixed* wavelength. The advantage of studying the wavelength-dependence is the much more detailed information that is made available. Since the exact spectral response is uniquely determined by the chemical composition of a sample, there are two distinct areas of application. Firstly, spectroscopy can be employed with pure substances for the purpose of obtaining more information on their molecular structure and other physicochemical properties; such are the research applications. Secondly, the characteristic nature of spectroscopic response can be utilised for the detection of particular chemical species in samples containing several different chemical components; these are the analytical applications. In both areas, lasers have made a very sizeable impact in recent years.

Although simple absorption and emission of light are the classic processes used for spectroscopic analysis, there are a great many other radiative interactions that are now being used for spectroscopic purposes. Even within the context of a single process such as absorption, there is a wide range of laser techniques that can be used for its detection. Also there are several inherently different classes of absorption spectroscopy, depending on the region of the electromagnetic spectrum being used. For example, molecular absorption spectra in the infra-red region generally result from vibrational transitions in the sample, and hence provide information on the structure of the nuclear framework, whereas absorption spectra in the visible or ultraviolet result from electronic transitions and so relate to electronic configurations. Whilst high intensity radiation can now be obtained from laser sources (in conjunction with frequency-conversion devices, if necessary) at virtually any wavelength longer than 100 nm, the crucial feature to consider for spectroscopic applications is the essential monochromaticity of the source.

In principle, laser monochromaticity can be used to very good effect, since the extremely narrow linewidth which can generally be obtained naturally lends itself to high-resolution spectroscopic techniques. Also, the small beam divergence facilitates the use of long path lengths through samples, which in the case of samples with very weak spectral features can helpfully improve the sensitivity. Ultimately, the resolution in any form of spectroscopy depends on the linewidths both of the radiation and the sample. The linewidth of a laser source is determined by various factors already considered in Sect. 1.5.3, such as natural line-broadening, collision broadening and Doppler broadening, and the same processes can contribute to broadening of spectroscopic features in a sample. However, there are two other line-broadening mechanisms which may come into effect in a sample in a particularly significant way due to the unique properties of laser light. These are *time-of-flight broadening*, and *power-* or *saturation-broadening*.

The former, applicable only to fluid media, results from the fact that molecules moving across a narrow laser beam with a perpendicular component of velocity v experience the radiation for only a very short period of time given by d/v, where d is the beam diameter. The result is the introduction of a frequency uncertainty of the order v/d. This effect is of course far more significant in gases than in liquids. Saturation broadening, on the other hand, is a reflection of the high intensities often associated with laser light. When the radiation has the appropriate frequency to induce molecular transitions (as of course is necessary for spectroscopic purposes) it can effect a considerable redistribution of population amongst the molecular energy levels. In other words, there is a departure from the conditions of thermal equilibrium under which the Boltzmann distribution applies. The intensity of absorption can thus decrease due to a reduction in the number of molecules left in the ground state; this is what is meant by *saturation*. Since saturation is most effective at the centre of an absorption band, the lineshape associated with the transition again broadens. Neither of these effects, however, alters the fact that with their inherently narrow linewidth, lasers offer the best spectroscopic resolution available.

Finally, there are certain other practicalities to consider when dealing with a monochromatic source. Many of the general instrumental features such as ratiometric recording have been discussed in the last chapter (Sect. 3.5.1). The primary consideration is that since any individual laser can only be operated over a limited range of the electromagnetic spectrum, the spectroscopic analysis of a particular sample may call for a particular laser system. Again, the monochromatic emission of a laser is such that conventional spectrometers operating with continuum radiation and monochromators are not directly applicable. Moreover, the traditional light sources often have the compensating virtues of greater reliability, simplicity of operation and more constant output. For these reasons, it has often tended to be the newer forms of spectroscopy which have been specifically developed around laser instrumentation. However, lasers have also made substantial inroads into all more traditional areas of spectroscopy by virtue of their high resolution capabilities. We begin with a consideration of some of the methods which are used to record absorption spectra using lasers.

4.1
Absorption Spectroscopy

Absorption spectroscopy is based on the selectivity of the wavelengths of light absorbed by different chemical compounds and involves monitoring the variation in the intensity of absorption from a beam of light as a function of its wavelength. The selectivity over absorption results from the requirement that photons absorbed have the appropriate energy to produce transitions to states of higher energy in the atoms or molecules of which the samples are composed (in the case of crystalline solids, more delocalised transitions involving the entire lattice can also take place). The strength of these transitions is governed by spectroscopic *selection rules*, each such process usually requiring the absorption of a single photon – though in Sects. 4.6 and 4.7 and Chap. 5 we shall encounter processes where this is not the case. Before proceeding further, we briefly recap on the basis equations for the absorption of light, for later reference.

To begin, it is clear that the rate of loss of intensity in a beam of a given wavelength passing through a medium absorbing at that wavelength is proportional both to the instantaneous intensity and the concentration of the absorbing species – thus we have the relation:

$$-\mathrm{d}I/\mathrm{d}l = \alpha I C, \tag{4.1}$$

where α is the constant of proportionality known as the *absorption coefficient*, l represents the path length through the sample, and C the concentration of the absorbing species. The absorption process generally promotes sample molecules to states which no longer absorb the incident radiation – this effectively leads to a small diminution of C and a corresponding reduction in the observed absorption. However, this *saturation* effect is only significant under exceptional circumstances (see, for example, Sect. 4.2.9). In most cases the sample concentration can be regarded as constant, and the solution to the simple differential equation (4.1) is the exponential decay function

$$I = I_0 e^{-\alpha I C}. \tag{4.2}$$

This result is generally known as the *Beer-Lambert Law* and expresses the intensity of light transmitted through the sample in terms of the intensity I_0 incident upon it. In order to facilitate expressions based on logarithms to base 10, Eq. (4.2) may alternatively be written as

$$I = 10^{-\varepsilon lC} I_0,$$ (4.3)

where $\varepsilon = \alpha/2.303$ is generally termed the *extinction coefficient* or the *molar absorption coefficient*. The product εlC is often given the symbol A and referred to as the *absorbance* or *optical density* of the sample; it is evidently related to the transmittance $T = I/I_0$ by

$$A = -\log_{10} T.$$ (4.4)

One final relation which follows from (4.3) is

$$\Delta I/I_0 = 1 - 10^{-A},$$ (4.5)

where $\Delta I = I_0 - I$; the left-hand side of Eq. (4.5) thus represents the fractional loss in intensity at a given wavelength on passage through the sample. This last relation will be useful for considerations of the sensitivity of the various methods for absorption spectroscopy.

Before the advent of the laser, most absorption spectroscopy was performed using broadband or continuum sources, with wavelength scanning carried out by passing the light through a monochromator placed either before or after the sample, as illustrated in Fig. 4.1a. In such a setup, the spectrum is obtained by detecting the intensity of radiation after passage through the sample, and plotting it as a function of the wavelength allowed through the monochromator. It is still the case that the majority of infra-red and uv/visible spectroscopic equipment in chemical laboratories operates on these principles, using conventional light rather than lasers. For later reference, it is interesting to note that

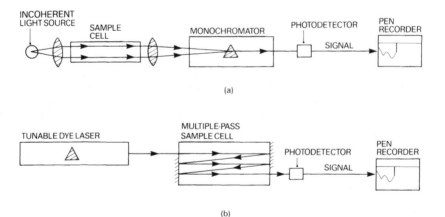

(a)

(b)

Fig. 4.1a and b Schematic diagrams of (**a**) a conventional absorption spectrometer and (**b**) the modified setup used with a tunable laser source. In each case, the signal received by the photodetector is ratioed with that obtained from a reference beam (not shown) derived from the same source, but which does not pass through the sample

for normal analytical purposes, absorbances of the order of 10^{-2} are considered the minimum which can be measured by conventional spectrophotometric techniques. Fourier transform instrumentation can offer an order of magnitude improvement on this figure.

Laser-based systems for absorption spectroscopy can be based either on fixed-frequency or tunable laser sources. Fixed-frequency lasers providing emission at only one, or a few discrete wavelengths, are not at all amenable to the usual absorption methods which require scanning over a continuous range of wavelengths, and thus call for specialised techniques. Two such methods, both developed in the late 1960s, are *laser magnetic resonance* and *laser Stark spectroscopy*, which are discussed later in this chapter. However, the introduction of tunable dye lasers at about the same time made it possible to obtain an absorption spectrum by scanning the source itself across the appropriate wavelength range. This kind of approach has many advantages and obviates the need for any monochromator: it is also a method which is applicable to other types of laser source, insofar as they can be tuned across the width of the gain profile (see Fig. 1.14). Although in this section we shall mostly concentrate on electronic spectroscopy in the visible range using a dye laser, the general principles discussed here are of much wider application and include infra-red spectroscopy with diode lasers.

We shall begin, then, by considering a simple setup for absorption spectroscopy using a dye laser, as illustrated in Fig. 4.1b. This arrangement produces a spectrum by monitoring the transmission through the sample as a function of wavelength; the attenuation relative to a reference beam provides a direct measure of the absorption, as in conventional spectrometry. However, absorbances as low as 10^{-5} can be detected in the laser configuration. Since the absorption signal is proportional to the distance travelled through the sample by the radiation, long path lengths through the sample are often used for laser spectroscopy; Fig. 4.1b illustrates one simple means of obtaining a long path length by multiple passes of the radiation through the sample medium. With a cell 20 cm in length, for example, it is fairly straightforward to generate 50 traversals, resulting in an effective path length of 10 m. In this way, an absorption band producing a decrease in intensity per traversal of as little as one part in 10^4 ultimately produces a drop in intensity of 0.5%, which is easily measurable. The technique is, of course, particularly well suited to studies of gases, which have very low absorbance; however, it is also of use in obtaining the spectra of components in dilute solutions, in which absorption features due to the solute have to be distinguished from the often very much more intense features due to the solvent. In such a case, the reference beam would be passed through a multiple-pass cell containing pure solvent. An alternative method of obtaining long path lengths is to use long hollow glass or quartz fibres filled with sample fluid.

Another method used to increase the absorption signal from samples is *intracavity enhancement*. As the name indicates, this phenomenon relates to the apparent increase in the intensity of absorption displayed by samples placed within the laser cavity. The detailed mechanism for this effect is somewhat complicated, since the losses resulting from sample absorption become an integral part of the laser dynamics. However, three effects contribute to the enhancement. Firstly, there is simply a multiple-pass effect resulting from the propagation of radiation back and forth within the cavity. Secondly, if the laser is operated close to threshold, then the introduction of extra losses can result in a drop to below threshold, so reducing the intensity of laser emission. The emission intensity is a very sensitive function of the losses close to threshold, so that even a very weak absorption can lead to a dramatic reduction in the output intensity. Finally, if the absorption linewidth of the sample is narrower than the laser gain bandwidth, competition between the various modes within the cavity can result in an effective transfer of energy to modes outside the absorption range, again reducing the emission intensity at the absorption wavelength. Hence, comparison of the intensity of laser emission with the sample, and then with a reference material placed within the cavity, provides a highly sensitive method for obtaining an absorption spectrum; in optimum cases, extinctions as low as 10^{-8} cm^{-1} can be detected. However, it should be noted that the mechanisms for the enhancement are such that there is no simple quantitative relationship between laser output and sample absorption.

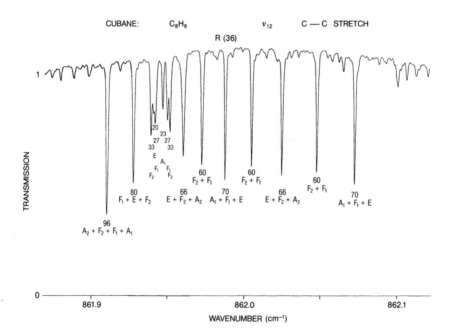

Fig. 4.2 Diode laser absorption spectrum of cubane obtained by Pine et al., showing the rotational structure of the n_{12} C-C stretch. Reprinted by permission from [1]. Copyright 1985, American Chemical Society

Finally, sensitivity can be improved by amplitude- or frequency-modulation of the incident light; here, the absorption signal is obtained by phase-sensitive detection of the oscillation in intensity of the transmitted light; usually the latter type of modulation is more successful. The application of such methods to near-infra-red absorption measurements based on diode lasers, in particular, offers ppb detection limits for a wide range of simple molecular species and as such finds many uses in trace gas detection and environmental monitoring. It is worth noting, however, that spectroscopically striking results can be obtained without any of the refinements described above. A very fine example is afforded by the infra-red absorption spectrum of cubane (C_8H_8) obtained by Pine et al. [1] shown in Fig. 4.2. This spectrum, obtained with a lead-salt diode laser, reveals the rotational fine structure in the C–C stretching band; notice that the entire scan covers a range of less than 0.2 cm^{-1}, and the wavenumber resolution is a remarkable 0.0004 cm^{-1}.

4.2
Specialised Absorption Techniques

Thus far, we have considered mostly direct methods of detecting absorption, involving measurement of the intensity of light after passage through the sample. As seen above, a common problem with such transmittance methods is the difficulty of detecting weak absorption features, since the signal ΔI is generally very small compared to I_0. There are, however, several alternative, but highly sensitive measurement techniques particularly suited to laser spectroscopy. These methods are all based on the monitoring of physical processes which take place *subsequent* to the absorption of radiation. Before examining these in detail, however, it is worth emphasising that nearly all the methods to be described in this section involve precisely the same absorption process in the initial excitation of the sample, and also produce spectra through analysis of the dependence on excitation frequency; hence they may all properly be described as types of absorption spectroscopy. There is one other type of absorption spectroscopy, entailing multiphoton absorption, which does not share the same excitation mechanism; discussion of this method is reserved for Sect. 4.6.

4.2.1
Excitation Spectroscopy

The deactivation of atoms and molecules excited by the absorption of visible or ultraviolet light can often involve the emission of light at some stage. In the case of atomic species, fluorescent emission generally takes place directly from the energy level populated by the excitation. In the case of molecular species, as illustrated in Fig. 2.16, p. 52, there are usually a number of different decay pathways which can be followed, of which spin-allowed fluorescence from the electronic state initially populated provides the most direct, and usually the

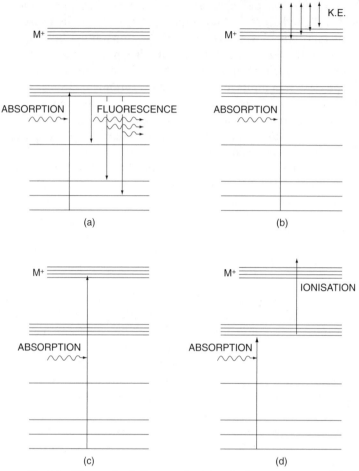

Fig. 4.3a-d Energy level diagrams for (**a**) excitation spectroscopy, (**b**) photoelectron spectroscopy, (**c**) zero kinetic energy photoelectron spectroscopy and (**d**) two-step ionisation spectroscopy of a simple species with discrete energy levels. The top levels denoted M$^+$ represent ion states

most rapid, means of deactivation. However, radiationless decay processes which occur in the vibrational levels prior to fluorescence result in emission over a range of wavelengths, so that even if the initial excitation is at a single fixed wavelength, the spectrum of the emitted light may itself contain a considerable amount of structure that can provide very useful information. This is the basis for fluorescence spectroscopy, discussed later in Sect. 4.3. *Excitation spectroscopy* or *laser-excited fluorescence*, by contrast, is concerned not with the spectral composition of the fluorescence, but with how the overall intensity of emission varies with the wavelength of excitation. Figure 4.3a illustrates the

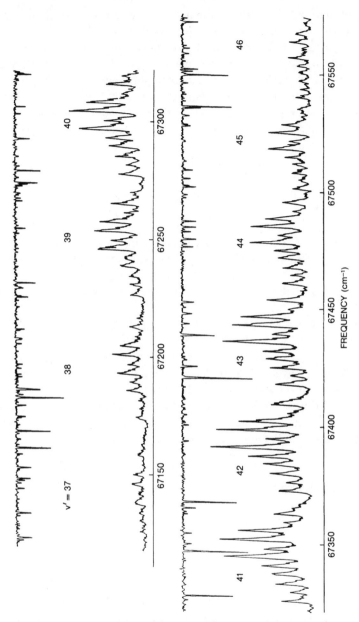

Fig. 4.4 Excitation spectrum of the B O_u^+ (v') ← X O_g^+ (O) system of Xe_2 obtained by Lipson et al. Reprinted with permission from [2]

various transitions giving rise to the net fluorescence from an atomic or simple molecular species with discrete energy levels.

The sensitivity of this method of absorption spectrometry stems from the fact that the signal is detected relative to a zero background; every photon collected by the detection system (usually a photomultiplier tube) has to arise from fluorescence in the sample and must thus result from an initial absorptive transition. If every photon absorbed results in a fluorescence photon being emitted, in other words if the *quantum yield* is unity, then in principle the excitation spectrum should accurately reflect both the positions and intensities of lines in the conventional absorption spectrum. In practice, collisional processes may lead to non-radiative decay pathways and thus a somewhat lower quantum yield, although this problem may to some extent be overcome in a gaseous sample by reducing the pressure. Also, the imperfect quantum efficiency of the detector has to be taken into consideration, along with the fact that a certain proportion of the fluorescence will not be received by the detector because it is emitted in the wrong direction. Nevertheless, it has been calculated that fractional extinctions $\Delta I/I_0$ as low as 10^{-14} are measurable by commercially available instrumentation using the method of excitation spectroscopy described above.

Amongst other applications, the high sensitivity of this method makes it especially well suited to the detection of short-lived chemical species, and Fig. 4.4 shows part of the spectrum of Xe_2 obtained by this method [2]. The vacuum ultraviolet radiation for the excitation was in this case produced by mixing the frequencies from two dye lasers pumped by a XeCl excimer laser source. Excitation spectroscopy is equally useful for identifying the transient species involved in gas-phase reactions, which represents one of its most important research applications (see Sect. 5.3.5).

4.2.2
Ionisation Spectroscopy

The second specialised technique used to monitor absorption in the uv/visible range is ionisation. The ions or free electrons produced by any such method can be detected by electrical methods with what is often close to 100% efficiency; ionisation spectroscopy thus affords one of the most sensitive methods for detecting absorption. As can be seen from Fig. 4.3b, direct laser photoionisation is not a technique directly amenable to spectroscopic application because ions can be produced on irradiation with any frequency above the ionisation threshold. With a fixed frequency input, it is nonetheless possible to resolve the various kinetic energies with which electrons escape from the nascent ion, and this is the basis of *ultraviolet photoelectron spectroscopy*. Resolution is generally poor, offering relatively little scope for characterisation of ion states. However, a relatively recent development illustrated in Fig. 4.3c offers a way around this problem. This is the technique known as *zero kinetic energy*

(ZEKE) photoelectron spectroscopy. Here, by tuning the input laser frequency and detecting only those electrons that trickle over the ionisation threshold with essentially zero energy, much more precise measurements can be made on the ion states, often with a resolution of around 1 cm^{-1}.

For the more usual spectroscopic study of neutral molecules, there is a great deal of merit in a two-stage process based on ionisation from states initially populated by the absorption of laser light, as illustrated in Fig. 4.3d. In this case, a regular absorption spectrum can be obtained by monitoring the rate of ion production as a function of the irradiation frequency. Various methods can be used to produce the ionisation, though clearly the process has to be sufficiently selective not to ionise ground state species. For excited states close in energy to the ionisation limit, ionisation can be induced either by application of an electric field, or by collisions with other atoms or molecules. Alternatively, a photoionisation technique can be employed, using either the laser or any other suitable frequency light source to produce photons with enough energy to bridge the gap to the ionisation continuum. Under ideal conditions virtually every atom or molecule excited by the laser radiation is then ionised and detected [3]. The simplest arrangement, in which single photons of laser radiation provide the entire (two-photon) energy for ionisation, falls under the general heading of multiphoton absorption and is discussed further in Sects. 4.6 and 4.7.

4.2.3
Thermal Lensing Spectroscopy

In thermal lensing spectroscopy, and also in photoacoustic spectroscopy discussed in the following section, the optical absorption from a laser beam is monitored through an effect based on the heating produced by absorption. The initial heating itself results from the radiationless decay of electronically excited states and may be termed a *photothermal* effect. In the case of thermal lensing spectroscopy, the specific phenomenon utilised is the temperature-dependence of the refractive index, which results in non-uniform refraction around the laser beam as it passes through any absorbing gas or liquid.

The mechanisms at work in this kind of spectroscopy, also known *as thermal blooming* spectroscopy, are in many respects more complicated than in any other absorption method. In the first place, the production of heat as the result of absorption depends on the detailed kinetics of relaxation in the sample molecules. The localised change in temperature which then ensues depends on the bulk heat capacity, and the extent of both conduction and convection in the sample; consequently, liquids are much more amenable to study by this method than gases. The characteristics of the laser beam also play a crucial role. With the spatial distribution of intensity across the laser beam normally conforming to a well-defined Gaussian profile, as shown in Fig. 4.5, the small volume of sample traversed by the centre of the beam experiences a greater

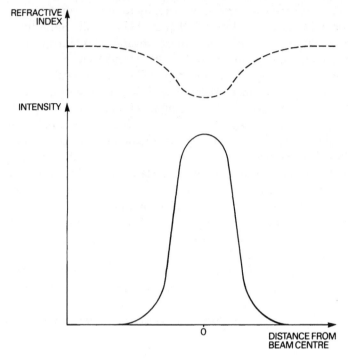

Fig. 4.5 The Gaussian variation in intensity of a laser beam across its beam (solid curve). The resulting variation in refractive index in a typical absorbing solid (dotted curve) produces the effect of convex lensing. In absorbing liquids the refractive index is reduced at beam centre, giving concave lensing

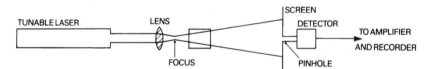

Fig. 4.6 The apparatus used in thermal lensing spectroscopy. The defocussing which arises as the laser is tuned through an absorption band of the sample results in a broadening of the transmitted beam and is observed as a decrease of intensity through the pinhole

intensity and thus exhibits a greater temperature rise than sample at the outside edge of the beam. The localised refractive index gradient that is consequently established is a function of the sample's thermal expansion properties and leads to a defocussing of the input laser beam (or of another probe beam) which is ultimately measured in the setup illustrated in Fig. 4.6. Sensitivity can be increased by chopping the beam and detecting the modulation in the signal detected.

The instrumentation required for thermal lensing spectrometry places several constraints both on the kind of laser which can be employed and on the nature of the sample that can be studied. It is a method more appropriate for absorption in the visible range than in the infra-red where direct heating effects occur, and it is generally implemented with a cw laser having a good transverse mode structure, so as to produce a consistent spatial intensity cross-section. Since the beam has to be focussed close to the sample in order for the defocussing to be observed, only a very small volume of sample is required to generate the spectrum, and the intensity of the signal does not increase in proportion to any increase in the path length.

Although this is a technique that can be applied for the characterisation of gases, it is mainly used for trace analysis in liquids. Here, samples must be carefully filtered to overcome false signals due to light scattering impurities; the method is basically inappropriate for flowing samples, unless the flow is very slow indeed (less than 1 ml per minute!) The dependence on the thermal properties of the sample make it a far more sensitive method for organic solvents than for water; in the former case, relative absorbances $\Delta I/I_0$ as low as 10^{-7} can be detected. This indicates that a detection limit as low as 10^{-11} M may be achievable for solutions of strongly coloured solutes. In this respect, it is worth noting that many analytical applications involve detection of a coloured derivative produced by complexation of the species to be detected. Here, the method often involves essentially colorimetric determination rather than true spectroscopy, though it is still useful to employ a tunable laser to optimise the wavelength for detecting absorption. Perhaps the most outstanding feature of thermal lensing spectroscopy is that it represents one of the cheapest means of obtaining high sensitivity in analytical absorption spectroscopy [4].

4.2.4
Photoacoustic Spectroscopy

As mentioned earlier this form of spectroscopy, also known as *optoacoustic* spectroscopy, is similar to thermal lensing spectroscopy in that both hinge on the heating effect produced by optical absorption. In this case, however, the specific effect which is measured results from the thermally produced pressure increase in the sample and makes use of the fact that if the laser radiation is modulated at an acoustic frequency, pressure waves of the same frequency are generated in the sample and can thus be detected by a piezoelectric detector or a microphone. Since the intensity of sound must depend on the amount of heating, it reflects the extent of absorption; hence, a spectrum can be obtained by plotting the sound level against the laser frequency or wavelength. Once again, the method produces spectra of very high resolution.

The usual instrumentation for laser photoacoustic spectroscopy is shown in Fig. 4.7. Visible or infra-red lasers can be used for the source, so that either

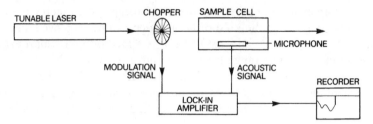

Fig. 4.7 The apparatus used in the photoacoustic spectroscopy of a gas

electronic or vibrational (or in some cases vibration-rotation) spectra can be obtained. The laser light first passes through a chopper which provides the acoustic frequency modulation of the beam incident upon the sample; a modulation signal is also derived at this point from a beam-splitter and optical detector (not shown on the diagram). The acoustic signal generated as the beam passes through the sample cell, or *spectraphone*, is then sent to the phase-sensitive amplifier locked into the modulation signal, and produces an output which is plotted by a pen recorder. Alternatively, a pulsed laser can be used in conjunction with an averaging system.

In contrast to thermal lensing spectrometry, photoacoustic absorption methods are easily applicable to both liquids and gases. Flowing fluids are also relatively easy to study, provided the acoustic noise generated by the flow is electronically filtered out of the microphone signal. In the case of liquids, it is again the case that solutions in organic solvents provide a much greater sensitivity than aqueous solutions, due to the high heat capacity of water. Nonetheless, it is important to choose a solvent that is comparatively transparent over the spectral range being examined, or else there may be little improvement over the spectra obtainable by conventional absorption methods.

Gases are the simplest to work with, and optimum sensitivity is obtained if the sample cell dimensions support resonant oscillations of the acoustic waves, which in practice means tailoring the modulation frequency to the particular sample being studied. Relative absorbances of 10^{-7} are measurable in both liquids and gases, although in gases the detection limit may be as much as two orders of magnitude lower. Sensitivity is in fact comparable to that of a good gas chromatography–mass spectrometry system, and concentrations of gases in the ppb (parts per billion) range can be detected over a wide range of pressures, from several atmospheres down to 10^{-3} atm. The technique is thus particularly well suited to investigations of gaseous pollutants in the atmosphere, and there are few other absorption techniques to match photoacoustic spectroscopy in sensitivity.

Finally, mention should be made of applications to solid samples. For chemically pure, relatively dense samples a common method of collecting the spectrum involves picking up the acoustic signal from a microphone placed in the gas above the sample. However in other cases, as for example in certain medical applications, it is possible to employ direct physical contact. Mostly such methods employ chemometric techniques based on photoacoustic response to near-infra-red radiation, using a sensor array incorporating semiconductor diode lasers. Such techniques now offer a rapid and non-intrusive method of determining tissue characteristics such as blood glucose levels.

4.2.5
Optogalvanic Spectroscopy

The last method commonly used to measure absorption spectra with a tunable laser source is based on the *optogalvanic effect*, which is essentially a change in the electrical properties of a gas discharge (or in some cases a flame) when irradiated by certain frequencies of light. In contrast to the other specialised methods discussed so far, this type of spectroscopy provides information on atomic or ionic species, rather than molecules. Also, the optical transitions involved are not only those originating from the ground state, since excited states are also appreciably populated.

The electrical current passing through a gas discharge results in ionisation of the atoms, and the efficiency of this process depends on the energy they already possess, being most effective for atoms in highly excited states near to the ionisation continuum. The current is thus a sensitive function of the various atomic energy level populations. In the presence of a beam of light with suitable frequency, absorption processes result in atomic transitions to states of higher energy, so that the relative populations of the various levels are changed. This effect can be registered as a change in the voltage across the discharge, which may be either positive or negative. Plotting the variation in discharge voltage against the irradiation frequency thus provides a novel kind of absorption spectrum and completely obviates the need for an optical detection system.

The apparatus used in discharge photogalvanic spectroscopy is shown in Fig. 4.8. In practice, the voltage change across an external resistor is monitored as a dye laser focussed into the discharge in a hollow cathode lamp is tuned across its spectral range. To increase the sensitivity, the laser radiation may

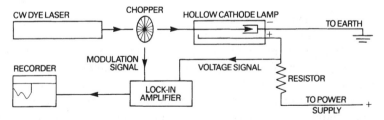

Fig. 4.8 Apparatus used for optogalvanic spectroscopy in a gas discharge

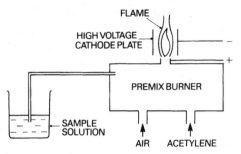

Fig. 4.9 Modified burner used for flame opto-galvanic measurements in laser-enhanced ioni-sation spectroscopy

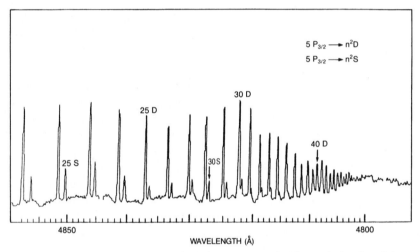

Fig. 4.10 Optogalvanic spectrum of rubidium. Reprinted with permission from [5]

again be chopped, and the modulation of the signal at the chopping frequency detected with a lock-in amplifier and recorder. By using a lamp containing uranium or thorium, whose spectra are already known with high precision, optogalvanic spectroscopy can incidentally be used as a method of wavelength calibration for tunable laser sources.

The primary analytical applications of optogalvanic spectroscopy arise elsewhere, however, in flame methods similar to those already widely used in atomic absorption and fluorescence spectrometry. Samples are usually in solution form and are introduced as a narrow jet into flames of an ethyne (acetylene)/air mixture. However, the optogalvanic instrumentation additionally requires the coupling of the burner head to the positive side of a power supply and the insertion of high-voltage cathode plates on either side of the flame, as shown in Fig. 4.9. The remainder of the setup is then similar to that shown in Fig. 4.8, with the burner assembly and cathode plates replacing the hollow cathode lamp. One alternative arrangement is to replace the cw dye laser with one which is flashlamp-pumped, using a photodiode to detect each pulse and trigger a signal-averaging amplifier.

In flames, it is generally found that the optogalvanic effect results in an *increase* in the extent of ionisation between the electrodes each time an absorption line is encountered. For this reason, the alternative designation *laser-enhanced ionisation* spectroscopy has also gained usage. Compared to the optical detection methods used in conventional flame spectrometry, this method suffers none of the problems usually caused by the flame background and scattering of the laser light. It also has improved sensitivity, both in spectral resolution and in its detection limits for certain elements. The best example of this sensitivity is its use in the analysis for lithium, which has been detected in concentrations as low as 1 pg/ml (10^{-12} g/ml) by this method, representing an improvement over the detection limit for most flame analytical methods by several powers of ten. The optogalvanic spectrum of rubidium is shown in Fig. 4.10.

4.2.6
Laser Magnetic Resonance

We now come to the first of the two methods mentioned earlier which enable absorption spectra to be obtained with fixed-frequency laser sources. Both of these, laser magnetic resonance and laser Stark spectroscopy, operate on the principle of tuning the *absorption* frequency of the sample to the frequency of the light source, rather than the more usual converse; the difference between the two is that one method involves tuning using a magnetic field, and the other an electric field. Both methods were contrived before tunable lasers had become commonplace and facilitated the various methods described above; however, they still have a more limited role to play in cases where high intensities are required, and only fixed-frequency laser sources provide sufficient power near to the absorption region of interest. For this reason, these two methods are often used for vibrational analysis in the low-frequency 'fingerprint' region below 1000 cm^{-1}, for example, using a carbon dioxide laser and also for obtaining rotational spectra in the far infra-red, using for example a hydrogen cyanide laser (wavelength 330 μm).

Laser magnetic resonance spectroscopy is based on the phenomenon known as the *Zeeman effect*; indeed, it is sometimes referred to as *laser Zeeman spectroscopy* and is generally applicable to molecules which possess a permanent magnetic dipole moment. Such a moment may result from various types of angular momentum, i.e. electron spin, electronic orbital angular momentum, nuclear spin, or even molecular rotation. However, the largest effects result from electron spin and are hence associated with paramagnetic species

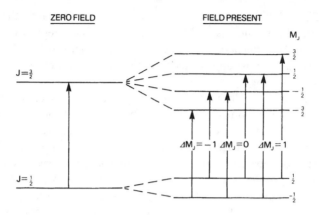

Fig. 4.11 Splitting of the J = $^1/_2$ and J = $^3/_2$ energy levels of a paramagnetic species in the presence of a static magnetic field, showing the allowed transitions

which have one or more unpaired electrons. States with a quantum angular momentum J usually have a (2J + 1)-fold degeneracy associated with the range of values of the azimuthal quantum number $M_J = -J$ through zero to $M_J = J$. Application of a static magnetic field results in each of these (2J + 1) levels being shifted in energy to a different extent, according to the equation

$$\Delta E = g\mu_B M_J B, \tag{4.6}$$

where g is the Landé factor determined by the kinds of angular momentum involved, μ_B is the Bohr magneton, and B is the magnetic induction field. Thus, for example, a J = $\frac{1}{2}$ → J = $\frac{3}{2}$ transition is split into six components satisfying the selection rule $M_J = 0, \pm 1$, as shown in Fig. 4.11. This leads to six different absorption frequencies (or three, if the Landé g factors of the J = $\frac{1}{2}$ and J = $\frac{3}{2}$ states are equal). Since the extent of energy shift is proportional to the applied magnetic field (Eq. 4.6), it is possible to bring absorption lines into resonance with a fixed frequency laser by varying the magnetic field strength; the resonance can then be detected by absorption of the laser light. A magnet capable of producing a 2 T field, for example, provides a tuning range of the order of 1 cm^{-1}.

The instrumentation for laser magnetic resonance is shown in Fig. 4.12. Because the method is a highly sensitive one, it is often applied to samples in the gas phase, and the lasers used are also generally gas lasers. It is therefore most straightforward to adopt an intracavity configuration, which brings about the advantage of signal ehancement discussed in Sect. 4.1. The spectrum is thus obtained by plotting the laser output as a function of the intensity of the magnetic field applied across the sample. Even greater sensitivity can be obtained

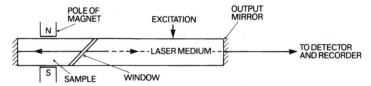

Fig. 4.12 Intracavity measurement of laser magnetic resonance

by modulating the magnetic field and using phase-sensitive detection methods, as in the related field of electron spin resonance spectroscopy; with such methods, it is estimated that a detection limit of 10^8 molecules cm^{-3} is achievable with current instruments.

Probably the most valuable attribute of laser magnetic resonance is its particular suitability to studies of short-lived paramagnetic species; for this purpose gas may be flowed through the sample cell immediately following the microwave irradiation or chemical reaction in which they are produced. The technique is thus very well suited to the monitoring of the free radicals involved in many photochemical reactions; the first high resolution gas phase spectrum of HO_2, for example, was obtained in this way. Laser magnetic resonance studies have also made possible the identification of an ethynyl (CCH) radical in the reaction of methane with free fluorine atoms [6]. Ethynyl, which is one of the most ubiquitous molecules in interstellar clouds, had not previously been detected on Earth.

4.2.7
Laser Stark Spectroscopy

In the Zeeman effect described above, a magnetic field removes the $(2J + 1)$-fold degeneracy from spin states with angular momentum J in molecules with a magnetic moment. In much the same way, an electric field lifts the degeneracy from rotational states in molecules possessing a permanent electric dipole moment; this is known as the Stark effect. In any polar compound, there is always a shift in energy which varies with the square of the electric field and is thus referred to as a *second-order*, or *quadratic* Stark effect. In symmetric rotor species, and in linear molecules in states with non-zero orbital angular momentum, there is an additional shift which is directly proportional to the field stength and is thus termed a *first-order*, or *linear* Stark effect. Again, the selection rule $\Delta M_J = 0, \pm 1$ applies to transitions between the various components of each rotational state. Thus, an absorption frequency associated with a single rotational transition in the absence of the field splits into several different absorption frequencies when the field is applied, and the exact frequency of each transition can be varied by changing the electric field strength; this is the basis of laser Stark spectroscopy.

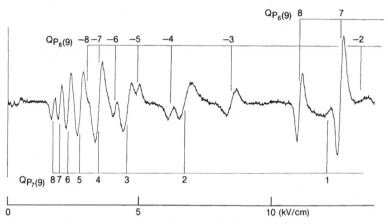

Fig. 4.13 Laser Stark spectrum of the ν_4 band of $CH_3C^{15}N$ obtained using the P(25) line from a CO_2 laser. Reprinted with permission from [7]

The source used for this kind of spectroscopy is usually a carbon monoxide or carbon dioxide laser, with apparatus basically similar to that shown in Fig. 4.12. The only main difference is that the poles of the magnet are removed, and electrode plates are inserted inside the sample cell so as to create a field transverse to the axis of the laser. Since an electric field of 10^4 cm^{-1} can produce a tuning range of only about 0.3 cm^{-1} even in a sample with a sizeable molecular dipole moment of 2 D, it is important to produce the highest possible field strength. As the magnitude of the field between the plates is inversely proportional to their separation, the optimum setup is where the plates are placed as close together as possible; however, the separation must remain sufficient for laser radiation to propagate back and forth in the cavity without interruption. Thus plate separations of a few millimetres are common, with applied voltages in the kilovolt range producing fields of the order of 10^4–10^5 V cm^{-1}. In cases where this is insufficient, it may be necessary to use more conventional extracavity methods so that the plates can be placed even more closely together. Sensitivity can again be improved by modulation of the applied field, coupled with phase-sensitive detection.

The method of laser Stark spectroscopy is rather more widely useful than laser magnetic resonance, since it is limited only to dipolar molecules. It is also sufficiently quantitative to enable very precise determinations to be made of molecular dipole moments. Its main use, however, lies in providing high resolution molecular rotation spectra in regions normally inaccessible by the traditional microwave methods. Figure 4.13 shows the rotational structure in the laser Stark spectrum of the ν_4 band of $CH_3C^{15}N$ obtained by Mito et al. [7].

4.2.8
Supersonic Jet Spectroscopy

One increasingly important method for achieving in practice the very high resolution that laser sources can afford is jet spectroscopy. This is a technique principally designed to exploit the well-known Joule-Thomson cooling effect, associated with the expansion of any gas through a small orifice into a region of lower pressure. When applied to polyatomic molecules in the vapour phase, the degree of cooling that can be achieved through supersonic expansion is sufficiently dramatic to effectively place each molecule in its lowest vibronic energy level. As a result, the numerous hot bands (transitions from thermally populated levels) that normally complicate and ultimately blur out spectral detail disappear, producing a clean and remarkably well-resolved spectrum. Jet expansion also has the effect of narrowing the velocity distribution, so that by measuring photoabsorption from a laser beam intersecting it at right angles, the effects of collisional and Doppler broadening are obviated in much the same way as by use of a molecular beam (see Sect. 5.3.5).

A splendid example of the art is illustrated in Fig. 4.14, giving a typical comparison between absorption spectra obtained (a) from a standard temperature and pressure vapour, and (b) from a supersonic jet expansion in helium. Here the spectra are those of the hydrocarbon tolane (diphenyl acetylene), the absorption measurements being made by detection of the fluorescence excitation as described in Sect. 4.2.1. Not only are some broad bands

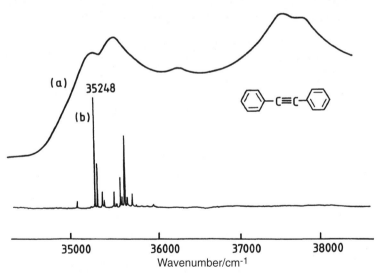

Fig. 4.14 Excitation spectra of diphenyl acetylene: (**a**) Normal vapour-phase spectrum; (**b**) Supersonic jet spectrum (cooling achieved by a modest four atmospheres excess pressure in a helium carrier gas). Reprinted with permission from [8]

completely removed by cooling (the jet having an effective vibrational temperature of a few degrees Kelvin) but a wealth of fine structure also emerges
from under the blanket. The associated sensitivity enhancement is a feature
that is increasingly a focus of interest for analytical applications.

4.2.9
Other High-Resolution Methods

Finally, we can briefly consider some other high-resolution absorption methods. One important method is based on detection of the resonant frequency ν_0
at the *centre* of a Doppler-broadened absorption band by *saturation spectroscopy*, for which the underlying principle is as follows. Consider an absorbing
sample strongly pumped by two counterpropagating laser beams of the same
frequency ν. If the laser frequency exactly coincides with the resonant frequency, then by virtue of the Doppler shift (Sect. 1.5.3) only molecules with
axial velocity component $v_k = 0$ can be excited. Consequently, if passage of
one beam through the sample is sufficient to appreciably deplete the ground
state population of these molecules, in other words to result in *saturation* taking place, then comparatively little light will be absorbed by the second beam.
This effect is often loosely referred to as *bleaching*. However, if the laser frequency is off-resonance by an amount $\Delta\nu$, then whilst one beam can excite
molecules with axial velocity $v_k = c\,\Delta\nu/\nu$, the beam travelling in the opposite
direction can excite other molecules with axial velocity $v_k = -c\,\Delta\nu/\nu$.

Thus the beam intensity is actually diminished less by absorption in the
former case, when its frequency coincides with the centre of the absorption

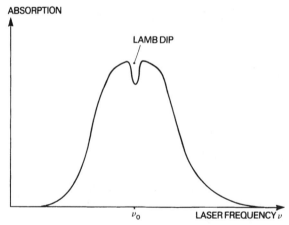

Fig. 4.15 Reduction of absorption intensity at the centre of
an optical absorption band observed with counterpropagating beams of the same frequency

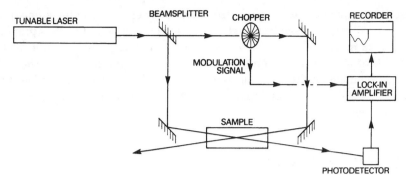

Fig. 4.16 Instrumentation for saturation Lamb-dip spectroscopy

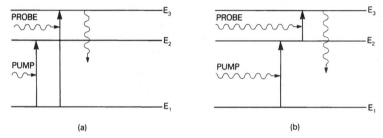

Fig. 4.17a and b Configuration of absorption transitions for optical-optical double resonance

band. This phenomenon is known as the *Lamb dip* and is illustrated in Fig. 4.15. In practice, the experimental setup for saturation spectroscopy usually involves two *nearly* collinear counterpropagating beams as shown in Fig. 4.16. The saturation technique is widely employed to eliminate Doppler-broadening, and its sensitivity can be improved by fluorescence detection. One further degree of sophistication can be introduced by separately chopping the two laser beams at different frequencies and observing fluorescence modulated at the sum frequency.

One final type of absorption spectroscopy based on specialised laser instrumentation is *optical–optical double resonance*, which calls for simultaneous irradiation of the sample by two laser beams of different frequency. There are various schemes for this method based on different combinations of optical transitions; the two which involve a pair of absorption transitions are illustrated in Fig. 4.17. In each case, however, the basic principle is the same; one transition $E_1 \rightarrow E_2$ is strongly driven by a modulated beam from a pump laser, and thus produces a fluctuation in the populations of the two energy levels it connects. A probe laser beam is then used to induce transitions originating from either level E_1, as in (a), or from E_2, as in (b). The fluctuating population of the initial state for the second transition thus results in a modulation of the probe beam intensity at the same frequency as the chopping of the pump laser. Phase-sensitive detection of this effect using a lock-in amplifier thus provides a sensitive means of monitoring absorption transitions and is useful for the assignment of high-resolution spectral lines in small molecules. Alternatively, use can be made of the fact that when the probe laser frequency strikes a resonance, it will result in diminished fluorescence from level E_2; this is the basis of *fluorescence dip spectroscopy*.

4.3
Fluorescence Spectroscopy

In every type of laser spectroscopy examined so far, the essential mechanism for introducing spectral discrimination has been the absorption of radiation by sample atoms or molecules. We now consider an alternative but equally well established branch of spectroscopy based on the emission of radiation by the sample. The laser provides a very selective means of populating excited states, and the study of the spectra of radiation emitted as these states decay is generally known as *laser-induced fluorescence*. There are two main areas of application for this technique: one is atomic fluorescence, and the other molecular fluorescence spectrometry.

4.3.1
Laser-Induced Atomic Fluorescence

This method is now increasingly being used for trace elemental analysis. It involves atomising a sample, which is usually in the form of a solution containing the substance for analysis, in a plasma, a furnace, or else in a flame, and subsequently exciting the free atoms and ions to states of higher energy using a laser source. Since the atomisation process populates a wide range of atomic energy levels in the first place, the emission spectrum is complex, despite the monochromaticity of the laser radiation. The fluorescence frequencies are nonetheless highly characteristic of the elements present, and measurement of relative line intensities can enable very accurate determination of flame temperature. Whilst detection capabilities are typically in the 10^{-11} g/ml range, concentrations two orders of magnitude smaller have been measured by this technique, corresponding to a molarity of 10^{-19} M. In fact, it has been demonstrated that laser-induced atomic fluorescence has the ultimate capability of detecting single atoms, although this is in an analytically rather unuseful setup based on a beam of identical atoms.

All sorts of lasers have been utilised for atomic fluorescence measurements. The most important factor to take into consideration is the provision of a high intensity of radiation in the range of absorption of the particular species of interest. For this reason, although individual measurements are made with a fixed irradiation wavelength, it can be helpful to employ a tunable source in

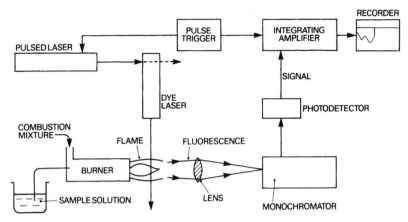

Fig. 4.18 The instrumentation for laser-induced atomic fluorescence measurements using a dye laser pumped with a pulsed laser and with flame atomisation of the sample solution

order to provide a detection facility for more than one element. For example, use of a frequency-doubled dye laser in conjunction with the traditional air/acetylene flame has been shown to offer dramatically improved detection limits for several precious metals [9]. In common with several other types of spectroscopy considered previously, sensitivity is often increased by either modulating a cw laser with a chopper, detecting the signal with a phase-sensitive lock-in amplifier, or else by integrating the signal using a pulsed laser, as shown in Fig. 4.18. The principles of laser-induced atomic fluorescence may also be employed in the emission spectroscopy of plasma, where the sample is introduced by laser *ablation*. Here, material is vapourised from the surface of a solid sample by a pulsed laser, and its elemental composition is characterised by its atomic fluorescence. This method is well suited to the direct determination of metals in solid samples, with detection limits typically in the ppm range. For heterogeneous solids, the microprobe principles discussed in Sect. 3.9 can be employed to good effect.

One other variation on this theme is *laser-induced breakdown spectroscopy*. Here, high intensity pulses of laser light, as for example from a Nd:YAG laser, are focussed into the sample and result in the formation of a spark through the process of dielectric breakdown. Once again, elemental analysis can then be carried out by measurement of the resultant atomic fluorescence. This method, which has evolved from the field of electric spark spectroscopy, has the advantage of obviating the need for electrodes, and hence removes any possibility of spectral interference associated with atomisation of the electrode surface. It also has the advantage of speed over many other analytical techniques since it requires little or no sample preparation. This kind of procedure enables ppb concentrations to be detected in optimal cases. Applications to tissue analysis and the identification of trace constituents of blood and sweat have been demonstrated, making the technique an attractive alternative to many of the methods more traditionally used in medical and forensic laboratories; field portable instruments for the analysis of metals in soils have also recently been developed.

4.3.2
Laser-Induced Molecular Fluorescence

Compared to atomic fluorescence, this method suffers from much poorer sensitivity due to the much broader lines in the emission spectra of all but the smallest polyatomic molecules. These result from the complicated radiationless decay processes involving vibrational energy levels, as discussed in the context of dye lasers in Sect. 2.5. The width of the emission bands can be reduced if the frequency of intermolecular collisions and other line-broadening interactions is diminished; for this purpose chemically simple samples may for example be analysed in the gas phase, or in the frozen matrix of another compound, as we shall see. However, the effects of reduced concentration associated with all such methods also have to be borne in mind.

In general, then, the high monochromaticity and narrow linewidth of a laser source does not result in a comparable resolution in laser-excited molecular emission spectra. However, it does present some advantages over the more traditionally used broadband radiation sources. Fluorescence spectra are often complicated by features which owe nothing to fluorescence but rather to Raman scattering in the sample, which results in the appearance of frequencies shifted by discrete amounts from the irradiation frequency (see the following section). The principal advantage of laser-induced fluorescence measurements is that when monochromatic light is used, the Raman features occur at discrete frequencies and can easily be distinguished from the broad fluorescence bands. It is also worth noting that for certain analytical applications fluorescence measurement may be more selective than absorption spectroscopy, since even when two different compounds absorb at the same frequency, their fluorescence emission frequencies may be quite different.

The energetics of molecular excitation and decay are such that fluorescence spectra can only be collected at wavelengths longer than the irradiation wavelength; subject to this restriction, however, there is comparative freedom over the choice of laser wavelength. Only in the spectroscopy of diatomic and other small molecules does the facility for selectively exciting a particular state play a significant role. However, fluorescence spectroscopy is not solely concerned with chemically stable systems; many important applications concern study of the photolytic processes which can occur in a sample through the interaction with laser light, as will be discussed in more detail in Chap. 5. When the energy absorbed by sample molecules is sufficient, fragmentation processes can occur, resulting in the formation of short-lived transient species which are often strongly fluorescent. For example, laser photolysis of cyanogen C_2N_2 produces free CN radicals which can be detected through collection of the fluorescence spectrum.

Another very useful feature of laser-induced molecular fluorescence is that with a pulsed source, it is possible to monitor the time-development of the emission process. This incidentally provides a further means of discriminating against the Raman scattering which takes place only during the period of irradiation, since the fluorescence is associated with decay processes subsequent to absorption. Different radiative transitions can in fact be discriminated by their different decay constants in some cases. Perhaps more useful in connection with the study of photochemical reactions, however, is the possibility of measuring excited state and transient lifetimes by this method, using either a Q-switched or a mode-locked source. With femtosecond pulses, it has now even become possible to determine the ultrashort timescale over which bond fission can occur (see Sect. 5.3.4). The methods of time-resolved laser measurement are thus particularly pertinent to the study of chemical reaction kinetics.

The measurement of fluorescence lifetime is also increasingly important in the study of biological systems. The main principle here is the fact that excited state decay times are often strongly influenced by the local chemical environment of the fluorophore, and in particular the proximity of any other group or groups to which energy can non-radiatively transfer. An interesting medical application concerns the monitoring of tissue metabolism, for which measurement of the ratio of nicotinamide adenine dinucleotide (NAD) to its reduced form NADH is a sensitive indicator. Only the latter absorbs 340-nm radiation, resulting in fluorescence emission at 480-nm. Moreover, the fluorescence lifetime increases from 0.5 to 1.0 ns on binding to proteins. Fluorimetry thus offers a direct and convenient means of assessing tissue status and can even be used for imaging different regions within individual cells.

The use of laser excitation in the study of molecular fluorescence lends itself naturally to measurements of polarisation behaviour, whose diagnostic value is increasingly being recognised. With a plane polarised source of excitation, the induced fluorescence is, in general, partially polarised, even when the sample is a material randomly oriented on the microscopic scale. Measurements generally entail resolving the fluorescence into components polarised parallel and perpendicular to the input, $I_\parallel$ and $I_\perp$ respectively, essentially as shown in connection with the Raman polarisation studies later illustrated in Fig. 4.24. Two parameters are commonly employed to quantify the extent of the polarisation. One is the *degree of fluorescence polarisation*, P, defined by

$$P = \frac{I_\parallel - I_\perp}{I_\parallel + I_\perp}, \tag{4.7}$$

the other is the *polarisation anisotropy*, r

$$r = \frac{I_\parallel - I_\perp}{I_\parallel + 2I_\perp}, \tag{4.8}$$

the two obviously related by $r = 2P/(3 - P)$. Optimally each parameter can be measured to an accuracy of about ± 0.01.

If the same electronic transition at a given site is involved both in the absorption of laser light and the subsequent fluorescence, the values $P = 0.50$ and $r = 0.40$ result for isotropic media. However, it is not uncommon for two or more different chromophore sites, within a protein molecule for example, to be directly involved in fluorescence response. This occurs through an intramolecular radiationless process known as Förster energy transfer where energy rapidly channels from the site at which laser light is absorbed to another, which then fluoresces. Where such processes intervene between absorption and fluorescence, the fluorescence polarisation parameters acquire a time-dependence whose measurement using pulsed laser instrumentation can provide a powerful insight into the kinetics [10].

In general, both cw and pulsed lasers are used for laser-induced molecular fluorescence measurements. Helium-cadmium and argon lasers are the most popular cw sources; nitrogen lasers and nitrogen-pumped dye lasers are usually adopted for pulsed sources. As usual, the latter are generally operated with a signal-integrating detection system. With such a setup, detection limits for solution samples can be as low as 10^{-13} M, or 10^{-5} ppb. With comparatively complex samples in which one is analysing for a particular component in admixture with a large concentration of other species, a detection limit of 1 ppb is more realistic. In air, for example, a detection limit of 10 ppb has been established for formaldehyde in an experiment which provides accurate concentrations within 100 s and without the need for sample collection, water extraction or chemical treatment of any kind [11] (see Fig. 4.19).

Laser-induced fluorescence is increasingly being used for the analysis of eluant from liquid chromatography (Sect. 3.8). It is also possible to 'tag' suitable biological species with fluorescent compounds, so as to facilitate laser fluorimetric detection. Mention should also be made of the novel analytical method which, although not strictly spectroscopic, is nonetheless based on fluorescence and is known as *remote fibre fluorimetry* (see Sect. 3.9). Quite another field of application lies in the remote sensing of oil spills or dissolved organic matter at sea, using an airborne ultraviolet laser source and telescoping detection system to scan the surface of the water for the characteristic fluorescence. In a similar way the concentration and health of seaborne phytoplankton can be monitored through excitation of its chlorophyll content, using a dye laser or the second harmonic of a Nd:YAG laser. Finally, laser fluorescence spectroscopy is widely employed for combustion diagnostics, since it enables the concentrations and temperatures of transient species within a flame to be ascertained with a high degree of spatial resolution, without interfering with the gas flow or chemistry. Intermediates which have been detected

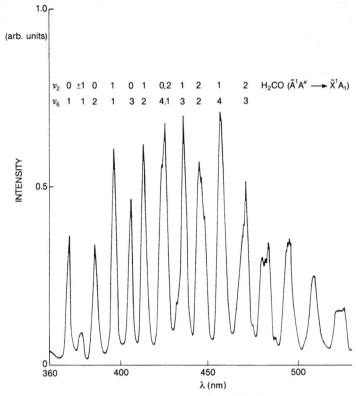

Fig. 4.19. Fluorescence spectrum of 2 mg l^{-1} formaldehyde in air, obtained by G.R. Mohlmann using a frequency-tripled Nd:YAG laser. The numbers denote the vibrational quanta in the electronic ground state, and the $\pm$ sign means that ν_2 is also excited in the electronically excited state. Reprinted with permission from [11]

by this method range from atomic oxygen and nitrogen to comparatively large fragments such as CH$_3$O, although OH is most commonly monitored since it provides a relatively straightforward measure of the degree of reaction.

Before leaving this subject, it is worth briefly taking a look at some of the ways in which the resolution of molecular fluorescence spectra can be improved in research on chemically pure samples. Of course the simplest method is to study the substance of interest in the gas phase, but this is rarely useful other than for small molecules with significant vapour pressure at low temperatures. One method makes use of a *supersonic molecular beam*, in which a jet of the sample mixed with helium is forced at high pressure through a small aperture into a vacuum. This process results in a beam of molecules in which there is only a very narrow distribution of translational energies, and only the lowest rotational states are populated. Hence, molecular collisions are comparatively infrequent, and both Doppler and rotational broadening of vibronic transitions are minimised, leading to a very significant improvement in spectral resolution. This kind of technique is particularly useful in the detailed study of reaction kinetics (see Sect. 5.3.5).

Other methods involve isolating individual molecules of the sample from each other in a low-temperature solid phase of some other substance. The solid phase may be constituted by a solvent in which the sample is dissolved and then frozen to liquid helium temperatures. Most commonly, however, the sample is vapourised and mixed with a large excess of an inert gas (typically 10^4–10^8 atoms per mole of sample), and then frozen onto a cold substrate in a vacuum line, a method known as *matrix-isolation*. In a sample prepared this way, even solute–solvent interactions are eliminated, and normally short-lived chemical transients can be studied since they are trapped in the matrix.

4.4
Raman Spectroscopy

Raman spectroscopy is the prime example of a laser technique that has been successfully developed from its specialised research origins to establish a significant presence in the modern analytical laboratory. Indeed, the process has recently been finding applications outside the lab, in on-line analysis and environmental monitoring. The technique is based on the *Raman effect*, a phenomenon involving the inelastic scattering of light by molecules (or atoms). The term 'inelastic' here denotes the fact that the scattering process results in either a gain or loss of energy by the molecules responsible, so that the frequency of the scattered light differs from that incident upon the sample. The energetics of the process are illustrated in Fig. 4.20. The two types of Raman transition, known as Stokes and anti-Stokes, are shown in Fig. 4.20a and b; the former results in an overall transition to a state of higher energy and the latter a transition to a state of lower energy. The elastic scattering processes repesented by Fig. 4.20c and d, in which the frequencies absorbed and emitted are equal, are known as *Rayleigh scattering*.

The first thing to notice about Raman scattering is that the absorption and emission take place together in one *concerted* process; there is no measurable time delay between the two events. The energy-time uncertainty relation $\Delta E \, \Delta t \geq h/2\pi$ thus allows for the process to take place even if there is no energy level to match the energy of the absorbed photon, since the absorption does not populate a physically meaningful intermediate state. It is therefore quite wrong to regard Raman spectroscopy as a fluorescence method, since fluorescence relates to an emissive transition between two physically identifiable states. In passing, we also note that despite their widespread adoption and utility, transition diagrams like Fig. 4.20 are strictly inappropriate for any such processes involving the *concerted* absorption and/or emission of more than one photon, since for example in this case they incorrectly imply that emission takes place *subsequent* to absorption.

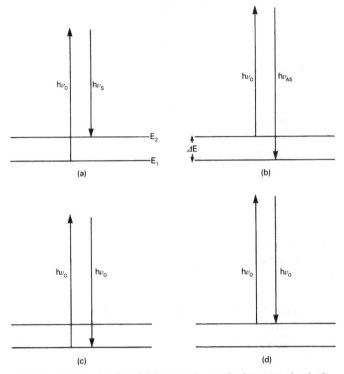

Fig. 4.20a-d Raman and Rayleigh scattering. Only the energy levels directly involved are depicted: (**a**) shows a Stokes Raman transition, and (**b**) an anti-Stokes Raman transition; (**c**) and (**d**) show Rayleigh scattering from the two different levels

Raman scattering generally involves transitions amongst energy levels that are separated by much less than the photon energy of the incident light. The two levels denoted by E_1 and E_2 in Fig. 4.20, for example, may be vibrational levels, whilst the energies of both the absorbed and emitted photons may well be in the visible range. Hence, the effect provides the facility for obtaining vibrational spectra using visible light, which has very useful implications as we shall see. In general, the Stokes Raman transition from level E_1 to E_2 results in scattering of a frequency given by

$$\nu_S = \nu_0 - \Delta E/h, \tag{4.9}$$

and the corresponding anti-Stokes transition from E_2 to E_1 produces a frequency

$$\nu_{AS} = \nu_0 + \Delta E/h, \tag{4.10}$$

where $\Delta E = E_2 - E_1$, and ν_0 is the irradiation frequency. For any allowed Raman transition, then, two new frequencies usually appear in the spectrum of

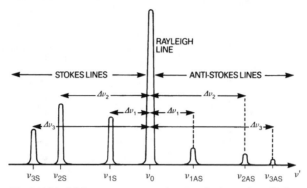

Fig. 4.21 Model Raman spectrum showing the intensity of light scattered with frequency v'

scattered light, shifted to the negative and positive sides of the dominant Rayleigh line by the same amount, $\Delta v = \Delta E/h$. For this reason Raman spectroscopy is always concerned with measurements of frequency *shifts*, rather than the absolute frequencies. In most cases, of course, a number of Raman transitions can take place involving various molecular energy levels, and the spectrum of scattered light thus contains a range of frequencies shifted away from the irradiation frequency. In the particular case of vibrational Raman transitions, these shifts can be identified with vibrational frequencies in the same way as absolute absorption frequencies in the infra-red spectrum.

One other point is worth raising before proceeding further. Although the Stokes and anti-Stokes lines in a Raman spectrum are equally separated from the Rayleigh line, they are not of equal intensity, as illustrated in Fig. 4.21. This is because the intensity of each transition is proportional to the population of the energy level from which the transition originates; the ratio of populations is given by the Boltzmann distribution (Eq. 1.13). There is also a fourth-power dependence on the scattering frequency, as the detailed theory shows. Hence the ratio of intensities of the Stokes line and its anti-Stokes partner in a Raman spectrum is given by

$$I_{AS}/I_S = \{(v_0 + \Delta v)/(v_0 - \Delta v)\}^4 (g_2/g_1) \exp(-h\Delta v/kT), \qquad (4.11)$$

and the anti-Stokes line is invariably weaker in intensity. The dependence of this ratio on the absolute temperature T can be made good use of in certain applications, for example in determining flame temperatures. However, since the Stokes and anti-Stokes lines give precisely the same information on molecular frequencies, it is usually only the stronger (Stokes) part of the spectrum that is recorded for routine analytical applications.

The Raman effect is a very weak phenomenon; typically only one incident photon in 10^7 produces a Raman transition, and hence observation of the effect calls for a very intense source of light. Since the effect is made manifest in

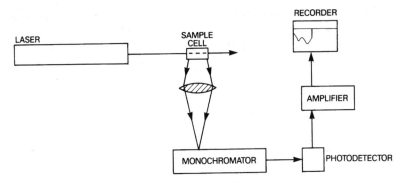

Fig. 4.22 Instrumentation for laser Raman spectroscopy

shifts of frequency away from that of the incident light, it also clearly requires use of a monochromatic source. Not surprisingly, then, the field of Raman spectroscopy was given an enormous boost by the arrival of the laser, so that Raman spectroscopy is today synonymous with *laser* Raman spectroscopy. A common setup is shown in Fig. 4.22. Light scattered by the sample from a laser source (usually an ion laser) is collected, usually at an angle of 90°, passed through a monochromator, and produces a signal from a photodetector. The Raman spectrum is then obtained by plotting the variation in this signal with the pass-frequency of the monochromator.

Even with a fairly powerful laser, Raman spectroscopy requires a very sensitive photodetection system capable of registering single photons to detect the weakest lines in the spectrum, and spectral acquisition can be a time-consuming process. Here, an alternative and more recently developed method of spectrum collection based on *array detectors* offers substantial advantages. With this kind of instrumentation, light scattered from the sample is first passed through a high-grade holographic filter to eliminate Rayleigh scattering, then dispersed and imaged onto a cooled charge-coupled device (CCD) array (see Sect. 3.4). In this way, the entire spectrum is collected rather than built up spectral element by element. The most obvious benefit is a significant improvement in the speed of data acquisition, a problem that had formerly always dogged Raman spectroscopy.

Raman transitions are governed by different selection rules from absorption or fluorescence. Thus whilst in centrosymmetric molecules only *ungerade* vibrations show up in the infra-red absorption spectrum, only *gerade* vibrations appear in the Raman spectrum. This illustrates the so-called *mutual exclusion rule*, applicable to all centrosymmetric molecules, which states that vibrations active in the infra-red spectrum are inactive in the Raman, and vice versa. Even for complex polyatomic molecules lacking much symmetry, the intensities of lines resulting from the same vibrational transition may be very different in the two types of spectrum, so that in general there is a useful complementarity between the two methods. Generally, it is the vibrations of the most polarisable groups which are strongest in the Raman spectrum, those of the most polar groups being strongest in the infra-red. A good illustration of this general complementarity is afforded by the spectra of toluene shown in Fig. 4.23. In the infra-red, for example, there is a very strong line at 726 cm^{-1}, almost absent in the Raman spectrum, which represents an all-in-phase out-of-plane CH mode of the aromatic system – the exact wavenumber

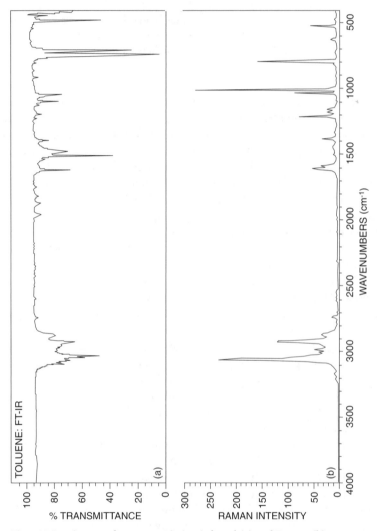

Fig. 4.23 Fourier-transform transmission infrared (**a**) and Raman (**b**) spectra of toluene, the wavenumbers in the Raman case as usual representing Stokes shift. In each case the resolution is 4 cm^{-1}. (By kind permission of Nicolet Instruments Ltd.)

suggesting either a mono or a 1,2-disubstituted ring. On the other hand, the strong Raman line at 1001 cm^{-1}, absent in the infra-red, is indicative of either mono, 1,3 or 1,3,5 substitution. Together, the spectra unequivocally identify the sample as a monosubstituted benzene. The alkyl nature of the substituent group shows up in the CH stretch bands at 2922 cm^{-1} in both spectra – in fact the pair of bands at around 1460 cm^{-1} and 1380 cm^{-1}, with opposite relative intensities in the two spectra, characterises the substituent as a methyl group.

In addition to providing vibrational frequency data, information can also be obtained on the symmetry properties of the vibrations themselves. This is accomplished by measurement of the *depolarisation ratios* of the lines in the Raman spectrum. The procedure is very simple; the sample is irradiated with plane polarised laser radiation with polarisation perpendicular to the plane in Fig. 4.22. The Raman spectrum is then collected with a polarising filter in-between the sample and monochromator, so as to analyse for the two perpendicular polarisation components of the scattered light, as shown in Fig. 4.24. The ratio of intensities in the spectra obtained from the two configurations, defined by

$$\rho_1 = I(z \rightarrow y)/I(z \rightarrow z), \tag{4.12}$$

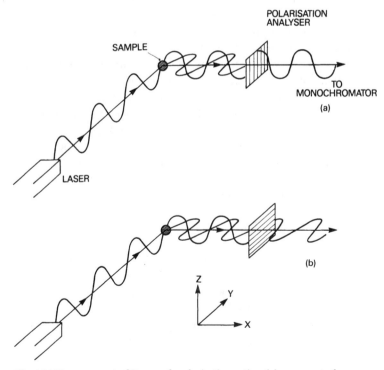

Fig. 4.24 Measurement of Raman depolarisation ratios; (a) represents the $(z \rightarrow z)$ configuration and (b) $(z \rightarrow y)$

can then be calculated for each Raman line and denotes the depolarisation ratio for the corresponding vibration. In the case of gases and liquids, ρ_1 takes the value of $\frac{3}{4}$ for vibrations that lower the molecular symmetry, but less than $\frac{3}{4}$ for vibrations that are totally symmetric.

All phases of matter are amenable to study by the Raman technique, and there are several advantages over infra-red vibrational spectroscopy in the particular case of aqueous solutions. Firstly, because only wavelengths in the visible region are involved, conventional glass optics and cells can be used. Secondly, since water itself produces a rather weak Raman signal, spectra of aqueous solutions are not swamped by the solvent. For these reasons, Raman spectroscopy is especially well suited to biological samples. As with most types of laser spectroscopy the spectrum is derived from a relatively small number of molecules because of the narrow laser beamwidth, and even with a comparatively insensitive spectrometer only a few millilitres of sample is sufficient. Raman spectroscopy is thus also a useful technique for the analysis of the products of reactions with low yield. One further point worth noting is that compared to infra-red spectroscopy, sample heating is much reduced by using visible radiation. The only case where heating does cause a real problem is with strongly coloured solids, where it has become a common practice to spin the sample so that no single spot is continuously irradiated; this is particularly important in the case of *resonance* Raman studies, discussed in Sect. 4.5.1.

One of the traditional problems with Raman spectroscopy has been interference in the spectrum of scattered light from sample fluorescence. There are various well-known ways to overcome this problem; for example, use can be made of the fact that the fluorescence signal will not generally exhibit any shift in frequency when the excitation frequency is changed, whereas of course the entire spectrum of Raman signals will shift by the same uniform amount. A more sophisticated method is to use pulsed lasers and time-gating techniques to separate the slower fluorescence from the spontaneous Raman emission. In some cases much simpler remedies are possible. In solids, for example, exposure to the laser radiation itself can substantially reduce the extent of fluorescence from defects and impurities – e.g. in industrial polymers about 80% of the fluorescence can disappear within a few seconds. In solutions it is sometimes expedient to add a fluorescence quencher with a known spectrum to accomplish the same result.

The most obvious solution, moving down to longer laser wavelengths where the photons have insufficient energy to induce fluorescence, at first sight appears unattractive because of the associated problem of increased levels of detector noise. However, the application of *Fourier-transform* techniques has neatly overcome these difficulties, enabling good fluorescence-free spectra to be collected using infra-red laser instrumentation. Commercially available Fourier-transform Raman instruments based on the Nd:YAG laser (1.064 μm) typically offer a speed of processing that can generate entire spectra within seconds [12]. However, there is a drawback: the trade-off against sensitivity

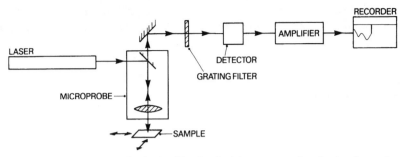

Fig. 4.25 Raman microprobe (point illumination) instrumentation. In the alternative global illumination configuration, a larger surface area is illuminated, and the detection equipment is replaced with an image intensifier phototube and camera

and power requirements. Dispersive instrumentation based on visible laser irradiation typically provides appreciably higher signal levels, as follows from the fourth-power dependence of Raman scattering intensity on emission frequency.

One type of application that is gaining popularity is known as the *laser Raman microprobe* and is principally used for heterogeneous solid samples, for example in corrosion studies and in geology. Two different methods have been developed based on the microprobe principles discussed in Sect. 3.9. In one method, illustrated in Fig. 4.25, various points on the surface of the sample (or indeed if it is transparent various regions within the sample) are irradiated with a laser beam and the Raman scattering is monitored, with a resolution of typically about 2 cm^{-1}. Since any particular chemical constituent should produce Raman scattering at one or more characteristic wavelengths, it is possible to filter out the emission from the surface at one particular wavelength and, hence, produce a map of the surface concentration of the substance of interest. Used in conjunction with a visual display unit and image-processing techniques, this method undoubtedly has enormous potential as a fast and non-destructive means of chemical analysis. A less sensitive, but simpler and more speedy technique involves global irradiation of the sample. In this way a small area of the surface can be imaged directly and the entire image passed through a filter to a camera system – here a workable resolution is commonly around 20 cm^{-1}.

4.5
Specialised Raman Techniques

As with absorption spectroscopy, there is a wide range of modifications to the standard methods of Raman spectroscopy described above. To some extent, these are modifications to the Raman process itself, rather than simply different means of detection; nonetheless, all the techniques discussed in this section involve essentially the same types of Raman *transition*. The one method which does not fit into this category, namely *hyper-Raman spectroscopy*, is discussed separately in Sect. 4.6.4.

4.5.1
Resonance Raman Spectroscopy

In Sect. 4.4, it was pointed out that the frequency of radiation used to induce Raman scattering need not in general equal an absorption frequency of the sample. Indeed, it is generally better that it does not, in order to avoid possible problems in accurately recording the Raman spectrum caused by the interference of absorption and subsequent fluorescence. However, there are certain special features which become apparent when an irradiation frequency is chosen close to a broad intense optical absorption band (such as one associated with a charge-transfer transition) which make the technique a useful one despite its drawbacks. Quite simply, the closer one approaches the resonance condition, the greater is the intensity of the Raman spectrum. The selection rules also change, so that certain transitions which are normally forbidden become allowed, thus providing extra information in the spectrum. Lastly, in the case of large polyatomic molecules in which any electronic absorption band may be due to localised absorption in a particular group called a *chromophore*, the vibrational Raman lines which experience the greatest amplification in intensity are those involving vibrations of nuclei close to the chromo-

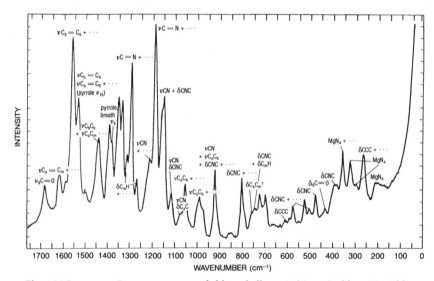

Fig. 4.26 Resonance Raman spectrum of chlorophyll a at 30°C, excited by a He-Cd laser at 441.6 nm (for the structure of chlorophyll *a*, see Fig. 5.6, p. 189). Reprinted from [13] by permission of John Wiley and Sons Ltd

phore responsible for the resonance. In principle, this facilitates deriving structural information concerning particular *sites* in large molecules. This feature has been much utilised, for example, in the resonance Raman vibrational spectroscopy of biological compounds containing strongly coloured groups and is well illustrated by the spectrum of chlorophyll *a* in Fig. 4.26. Also significant in this context is the possibility of using mode-locked laser excitation so as to derive structural data on a picosecond timescale. Here, resonance Raman spectra provide far more information than absorption spectra and facilitate time-resolved studies of ultrafast biological processes such as those involved in photosynthesis and vision (see Sect. 5.3.4).

Figure 4.27 shows the energetics of Raman scattering under different conditions with increasing irradiation frequency, with the conventional non-resonant process represented by (a). In the situation corresponding to (b), there is a sizeable *pre-resonance* growth in intensity even before the irradiation frequency reaches that of an absorption band. When the incident frequency is coincident with a discrete absorption frequency as in (c), real absorption and

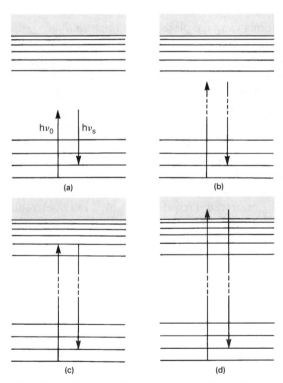

Fig. 4.27a-d Diagrams showing the same Stokes Raman transition with variation in the irradiation frequency ν_0: (**a**) conventional Raman effect; (**b**) pre-resonance Raman effect; (**c**) resonance fluorescence; (**d**) resonance Raman effect

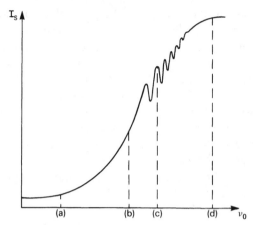

Fig. 4.28 Resonance Raman excitation profile of
the transition illustrated in Fig. 4.27

emission transitions can take place, and *resonance fluorescence* starts to complicate the picture. Nonetheless, it is when the incident photons have sufficient
energy to bridge the gap to a continuum state (d) that the largest Raman intensities are produced, often with an amplification factor of up to a thousand
over the normal Raman signal: this phenomenon is known as the *resonance
Raman effect*. If the intensity of a given Raman line is plotted as a function of
the irradiation frequency, then a graph like that shown in Fig. 4.28 is produced; this is known as a *resonance Raman excitation profile*.

Resonance Raman spectroscopy can be performed with any laser source
provided it emits a wavelength lying within a suitable broad absorption band
of the sample. Clearly, for application to a range of coloured samples, a tunable
dye laser is to be preferred. Frequency-doubled ion laser radiation is a better
choice for materials whose principal absorption is in the ultraviolet. Unfortunately, resonance fluorescence occurs with increasing intensity and over an
increasingly broad range of emission wavelengths as the resonance condition
is approached. For certain applications this problem may be overcome by use
of a mode-locked source providing ultrashort (picosecond) pulses. Since there
is no detectable time-delay for the appearance of the Raman signal, whilst resonance fluorescence is associated with a lifetime typically in the nanosecond
range, the signal from the photodetector can be electronically sampled at suitable intervals and processed so that only the true Raman emission is recorded
(see Sect. 3.5.2).

4.5.2
Stimulated Raman Spectroscopy

As has been stressed before, the Raman effect is essentially a very weak one, and it is in general only under conditions of resonance enhancement that the scattering intensity could be regarded as appreciable. However, there is one other way in which the effect can be enhanced, which is as follows. If a sufficiently intense laser is used to induce Raman scattering, then despite the initially low efficiency of conversion to Stokes frequencies, Stokes photons which are emitted can *stimulate* the emission of further Stokes photons through Raman scattering of laser light by other sample molecules in the beam. This process evidently has the self-amplifying character always associated with stimulated emission and is therefore most effective for Stokes scattering in approximately the direction of the laser beam through the sample. In fact with a giant pulse laser this *stimulated Raman scattering* can lead to generation of Stokes frequency radiation in the 'forward' direction with a conversion efficiency of about 50%.

Once a strong Stokes beam is established in the sample, of course, it can lead to further Raman scattering at frequency

$$\nu'_S = \nu_S - \Delta\nu, \tag{4.13}$$

and this too may be amplified to the point where it produces Raman scattering at frequency

$$\nu''_S = \nu'_S - \Delta\nu, \tag{4.14}$$

and so on. Hence a series of regularly spaced frequencies $(\nu_0 - m\Delta\nu)$ appears in the spectrum of the forward-scattered light, as illustrated in Fig. 4.29. Because of the self-amplifying nature of the process, it is usually the case that only the vibration producing the strongest Stokes line in the normal Raman spectrum is involved in stimulated Raman scattering. The effect does not, therefore, have the analytical utility of conventional Raman spectroscopy.

One other phenomenon generally accompanies stimulated Raman scattering and is not always clearly differentiated from it. This is a *four-wave mixing* process, whose energetics are illustrated in Fig. 4.30. It provides a mechanism for the conversion of two laser photons of frequency ν_0 into a Stokes and anti-Stokes pair, with frequencies $\nu_S = \nu_0 - \Delta\nu$, and $\nu_{AS} = \nu_0 + \Delta\nu$. Since this process returns the molecule in which it takes place to its initial state, there is no relaxation period necessary before two more laser photons can be converted in the same molecule, which of course significantly improves the conversion efficiency. There are other reasons why the process is particularly effective, however. One is the fact that if the photons involved propagate in a suitable direction, four-wave mixing can take place without any transfer of momentum to the sample; this means that the total momentum of the absorbed photons must equal that of the two emitted photons. As noted in Sect. 3.2.2, photons carry a

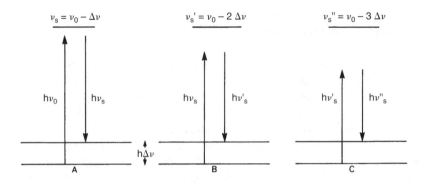

Fig. 4.29 Illustration of the stimulated Raman effect. As the laser photon of frequency ν_0 travels through the sample, it suffers three consecutive frequency conversions through identical Stokes Raman transitions in three different molecules A, B and C

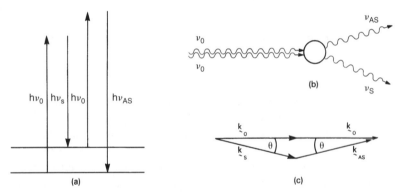

Fig. 4.30a–c Four-wave process involved in stimulated Raman scattering and CARS: (a) shows the energetics, (b) depicts the physics and (c) illustrates the wave-vector matching condition

momentum $hk/2\pi$ in the direction of propagation, where k is the wave-vector of magnitude $k = 2\pi\nu/c' = 2\pi n/\lambda$, and n is the refractive index of the medium. Because the refractive index generally depends on wavelength too, the required wave-vector matching is in this case generally produced when there are small angles between the wave-vectors as shown in Fig. 4.30b and c.

The four-wave process thus returns the molecule to its initial state and confers no momentum to it. Any mechanism which fulfils these two conditions is known in the field of non-linear optics as a *coherent parametric process* and is

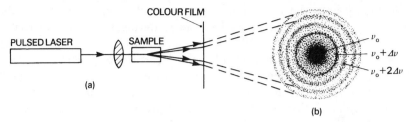

Fig. 4.31 Photographic demonstration of the four-wave stimulated Raman effect: (a) shows the apparatus, and (b) the photograph, assuming that the Stokes frequencies $\nu_0 - m\Delta\nu$ are outside the range of sensitivity of the photographic emulsion

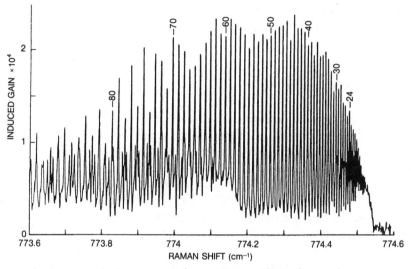

Fig. 4.32 Raman gain spectrum of SF_6 at 3.8 torr, obtained using the 647.1-nm line from a krypton ion laser as probe, showing ν_1 fundamental Q(J) transitions and underlying $\nu_1 + \nu_6 \leftarrow \nu_6$ hot band transitions. Reprinted from [14] by permission of John Wiley and Sons Ltd

invariably associated with a substantially increased conversion efficiency. Since in the experimental context described above the Stokes transition involved can also be a stimulated process, this four-wave interaction can therefore take place at a very significant rate, and it plays an important role in the production of further anti-Stokes frequencies $(\nu_0 + m\Delta\nu)$. The angle-dependence of the emission can be graphically illustrated by placing a colour-sensitive photographic film as shown in Fig. 4.31a, producing rings of colour as in (b). Most of the converted frequencies are emitted within an angle of about 10° away from the laser beam direction. The efficiency of this process in producing Stokes and anti-Stokes frequencies in liquids or highly pressurised gases is sufficiently high enough to make the effect useful as a means of laser frequency conversion (see Sect. 3.2.3).

Finally, we note that there is more than one other spectroscopic method based on the principle of stimulated Raman scattering. One of the alternatives is to simultaneously irradiate the sample with both a 'pump' laser beam of frequency ν_0 and with another tunable 'probe' laser beam. When the frequency of the latter beam is tuned through a Stokes frequency ν_S, it stimulates the corresponding Raman transition and thus experiences a gain in intensity. The Raman spectrum is thus obtained by plotting the intensity of the probe beam after passage through the sample against its frequency; this method is known as *Raman gain spectroscopy*. The beautiful example of a Raman gain spectrum shown in Fig. 4.32 illustrates the rotational fine structure in the fundamental breathing mode of SF_6. The other common method involving stimulated Raman transitions is inverse Raman spectroscopy, which has a rather different methodology as described below.

4.5.3
Inverse Raman Spectroscopy

The stimulated Raman processes discussed in the last Section involve the absorption of a frequency ν_0 and the stimulated emission of a Stokes frequency ν_S. *Inverse* Raman spectroscopy is based on the converse radiative processes, in other words absorption of a Stokes frequency (or for that matter an anti-Stokes frequency), and stimulated emission of ν_0, as shown in Fig. 4.33. In practice, it is usual to employ an intense monochromatic laser source for the pump beam of frequency ν_0, and a secondary *continuum* source of lower power to act as probe; inverse Raman transitions are then detected as absorptions from the probe beam at each Stokes or anti-Stokes frequency. Thus although it is possible to use a tunable dye laser pumped by the pump laser for the probe, the broad band fluorescence from the laser dye can in fact be used directly without being monochromatised, as shown in Fig. 4.34a. Note that the energetics of the inverse Raman effect are such that the *anti-Stokes* transition is normally the

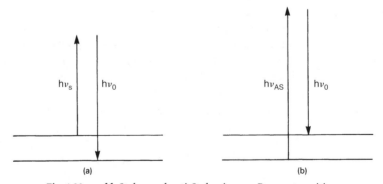

Fig. 4.33a and b Stokes and anti-Stokes inverse Raman transitions

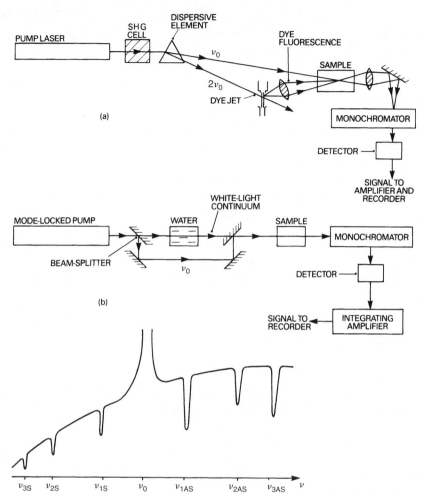

Fig. 4.34a - c Measurement of inverse Raman spectra: (**a**) using a laser dye to generate a continuum, and (**b**) using a white light continuum produced by self-focussing of picosecond laser pulses in water; (**c**) illustration of the spectrum (compare Fig. 4.21)

more intense since it originates from a state of lower energy, and hence with a higher population than does the Stokes transition. Thus the spectrum is best obtained with probe frequencies *higher* than the pump beam; this is the reason for the use of a frequency-doubling crystal in Fig. 4.34a.

An alternative method makes use of the broad-band *ultrafast superconti-nuum laser source* produced by focussing intense ultrashort laser pulses into an optically transparent but physically dense material (see Sect. 3.3.3). The physical mechanism for this process is complex, but the effect is delightfully simple to produce; focussing into a beaker of water will often suffice to produce a continuum covering the whole range of the visible spectrum. The setup used to obtain inverse Raman spectra using this source as the probe is shown in Fig. 4.34b, and c shows the kind of spectrum obtained by either inverse Raman method.

4.5.4
CARS Spectroscopy

Next we consider *coherent anti-Stokes Raman scattering* spectroscopy, usually abbreviated to CARS. On the molecular level, the mechanism for this process is precisely the same as the four-wave interaction encountered in Sect. 4.5.2 in connection with the stimulated Raman effect, and all of the illustrations in Fig. 4.30 are directly applicable to CARS. The difference arises in the method used to induce the effect. Whereas in the case of the stimulated Raman effect the Stokes wave is spontaneously generated by conventional Raman scattering from a beam of frequency ν_0, in CARS it is produced by a second laser beam directed into the sample, whose frequency ν_1 is tuned across the frequency range below ν_0. The four-wave interaction (see Fig. 4.35) produces coherent emission at a frequency

$$\nu' = 2\nu_0 - \nu_1, \tag{4.15}$$

so that when ν_1 coincides with *any* Stokes Raman frequency ν_S, we have

$$\nu' = 2\nu_0 - \nu_S = 2\nu_0 - (\nu_0 - \Delta\nu) = \nu_0 + \Delta\nu = \nu_{AS}, \tag{4.16}$$

corresponding with the anti-Stokes frequency. Since it is once again a coherent parametric process, the CARS emission is produced in a well-defined direction governed by the wave-vector matching condition as in Fig. 4.30 (c). As with stimulated Raman scattering, this directionality ensures that all the output can readily be collected for spectroscopic analysis. This contrasts with conventional Raman scattering, where only a small fraction of the global emission is normally collected.

In contrast to the stimulated Raman effect, emission at the frequency given by Eq. (4.15) can also occur, albeit with reduced intensity, when ν_1 does *not* equal a Stokes frequency. The distinction is illustrated in Fig. 4.35: (a) shows the *resonant* CARS mechanism (cf. Fig. 4.30a); (b) shows the *non-resonant* process that occurs when $\nu_1 \neq \nu_S$. Hence, although the strongest lines in the CARS spectrum occur when ν_1 strikes resonance with a Stokes frequency, this is usually seen against a background emission due to the relatively frequency-insensitive non-resonance mechanism. For this reason, CARS spectroscopy is not well suited to applications in trace analysis. However, since it does produce a spectrum typically 10^4–10^5 times more intense than the normal Raman effect, but subject to the same selection rules, there are other important areas of application, as we shall see.

Although CARS is a two-beam method basically requiring two laser sources, part of the output from a single pump laser may be frequency-converted in a

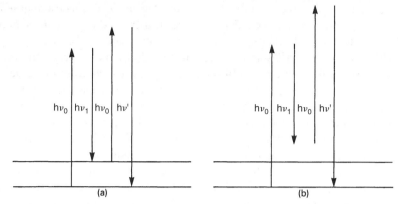

Fig. 4.35a and b (a) Resonant, and (b) non-resonant four-wave interactions in CARS spectroscopy

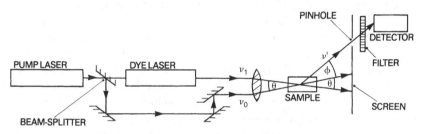

Fig. 4.36 Instrumentation for CARS spectroscopy

dye laser cavity to produce the second beam, as shown in Fig. 4.36. The two laser beams are directed into the sample at a small angle θ, and as the dye laser is tuned, CARS emission is detected through a pinprick hole in a direction making an angle ϕ with the ν_0 beam, as shown. In fact, the two angles θ and ϕ are nearly equal, since the wave-vector matching triangle of Fig. 4.30c is almost isosceles. In principle, since the CARS emission frequency is given by Eq. (4.15), it is completely determined by the two incident frequencies, and the spectrum may be obtained by plotting the intensity of light received by the detector against the frequency difference $\nu_0 - \nu_1$.

Occasionally, a monochromator may be placed before the detector to cut out stray frequencies resulting from light scattering and fluorescence processes in the sample. However, it is often sufficient to use a simple optical filter with a cut-off just above ν_0 for this purpose, making use of the fact that only higher frequencies can result from the CARS process.

An unusual feature of CARS, but one which it holds in common with other coherent parametric processes, is that it depends quadratically on the number of molecules per unit volume. Hence, whilst the vibrational spectra of liquids can be obtained with cw lasers, giant pulsed lasers producing megawatt pulses are required in order to obtain the vibration–rotation CARS spectra of gaseous samples; a frequency-doubled Nd:YAG source is a common choice. It is, none-theless, a most useful technique for samples obtainable only in very small quantities; since the CARS signal is produced only by molecules at the inter-section focus of the two applied beams, only microlitre volumes of liquid are required, or gas pressures of 10^{-6} atm. Two other points are worth noting con-cerning gaseous samples. Firstly, because the refractive index is close to unity for each frequency involved, the angles θ and ϕ in the wave-vector matching diagram are sufficiently small that in practice a collinear beam geometry can be used. Secondly, there is almost no Doppler broadening of spectral lines, since the effect is based on the absorption and emission of photons with al-most identical wave-vector. Thus CARS can be used to produce spectra of very high resolution, with bandwidths as low as 10^{-4} cm^{-1} in optimum cases.

One special area in which CARS spectroscopy has found widespread appli-cation, by virtue of its particular suitability to highly luminescent samples, is in combustion and other high-temperature reaction diagnostics. The main ad-vantage over traditional Raman spectroscopy is the fact that a coherent high-frequency output beam is produced, which can be much more easily detected against a strong fluorescence background. Here the CARS technique often pro-vides results which usefully complement the data on transients available from laser-induced fluorescence measurements. By making the two applied laser beams intersect within the flame or reaction volume to be studied, the CARS spectrum not only helps to reveal the chemical composition in any small re-gion, but also the temperature can be accurately determined from the relative intensities of rotational lines. This method of temperature determination not only surpasses that of any standard thermocouple device in its degree of accu-racy, but also in that measurements are possible well above the usual thermo-couple limit of about 2500 K. The technique is moreover non-instrusive, and hence in no way affects the process being studied. A good industrial example of this type of application lies in the analysis of gas streams in coal gasification plants; the technique has also been widely applied to studies of the internal combustion engine.

4.5.5
Surface-Enhanced Raman Spectroscopy

Another special type of Raman spectroscopy involves the study of surface-ad-sorbed species. In view of the low intensities normally expected in Raman spectra, the technique does not immediately seem a very sensible choice for the study of very low surface concentrations – indeed it is only quite recently that instrumental sensitivity has been developed sufficiently to allow interfa-cially specific Raman studies without necessary exploitation of the enhance-ment mechanism to be discussed below. However, for species adsorbed onto suitable metallic substrates (silver is the favourite), the Raman signal from an

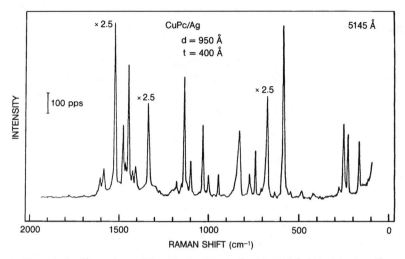

Fig. 4.37 Surface-enhanced Raman spectrum of copper phthalocyanine on silver obtained by Hayashi and Samejima using 514.5-nm radiation from an argon ion laser. Reprinted from [15] by permission of North Holland Physics Publishing

adsorbate is often found to be up to six orders of magnitude higher than might be expected. There has been a great deal of debate about the reason for this enhancement and the factors by which it is influenced, and it has become clear that more than one mechanism is at work. Surface roughness is one effect which seems to play an important role. It is also evident that it is the molecules closest to the metal surface whose Raman transitions experience the greatest enhancement. Figure 4.37 shows the surface-enhanced Raman spectrum of a 95-nm layer of copper phthalocyanine on an evaporated silver film substrate. The high luminescence level and low solubility of phthalocyanine compounds usually result in a comparatively poor signal-to-noise ratio in the Raman spectrum, but the enhanced thin-film spectrum gives a very good result. The enhancement factor here is approximately 20.

There is potentially a great deal of practical utility in the effect, and surface-enhanced Raman spectroscopy (SERS) has gained some popularity, albeit for certain rather specific applications – the enhancement tends to be very sensitive to the input wavelength, calling for careful choice of the laser line best suited for the substrate. One particular area of interest lies in solution studies using electrochemical cells with silver electrodes. Silver electrodes can be very effectively roughened by oxidation-reduction cycles, and solute species can subsequently be studied at the electrode surface by SERS. Symmetry considerations often permit the orientation of adsorbed molecules to be deduced by this means. SERS is now being considered for a number of chemical sensor applications; the process also shows potential for characterisation of the surface-adsorbed species involved in heterogeneous catalysis.

4.5.6
Raman Optical Activity

The last subject we shall deal with in this section concerns Raman scattering measurements on optically active compounds, usually in the liquid or solution state, using circularly polarised laser light. Using the same type of electro-optic modulation of polarisation discussed in Sects. 3.1 and 3.7, it is possible to obtain a spectrum showing the *difference* in the Raman intensity $I^R - I^L$ as a function of scattering frequency. In contrast to most conventional measurements of chirality such as optical rotatory dispersion or circular dichroism, which are based on molecular electronic properties, *circular differential Raman spectroscopy* is based on the molecular vibrations. For this reason it is more directly related to the detailed stereochemical structure ultimately responsible for any manifestation of chirality. The information contained in a circular differential Raman spectrum is also far more directly useful. In particular, the extent of differential scattering in a region of the spectrum associated with a particular group frequency can be interpreted in terms of the chiral environment of the corresponding functional group.

Although only relatively simple chiral species appeared amenable to study by means of Raman optical activity in its infancy, the technique has now matured to the point where it can generate highly useful structural information on complex biological structures such as enzymes and other proteins. Figure 4.38 shows the Raman circular intensity sum and difference spectra of lysozyme in aqueous solution, as recently obtained in the laboratory of Laurence Barron, a pioneer of the method. A number of important features can be identified in the difference spectrum. In the 850–1150 cm^{-1} region, bands generally originate from backbone skeletal stretch vibrations (C_α–C and C_α–N) and so afford good signatures of secondary structure. In the case of lysozyme, positive intensity features in the range 880–960 cm^{-1} can be assigned to α-helix, others in the range 1000–1060 cm^{-1} to β-sheet elements of structure. Secondary structure is also manifest in the 1200–1400 cm^{-1} region, where amide III modes involving coupled C_α–H and N–H deformations occur. More importantly, however, this region also contains bands characteristic of loops and turns, such as the negative intensity feature at 1240 cm^{-1} in the difference spectrum of lysozyme. The identification of such structural elements which characterise the overall three-dimensional structure (the tertiary fold) of proteins in solution is a signal advantage over conventional Raman spectroscopy [16]. This also allows the technique to be applied to the study of protein folding and other dynamical problems.

The only other method which has the potential to offer data of this kind is vibrational circular dichroism using infra-red radiation, a technique which is very insensitive for low-frequency vibrations. Circular differential Raman scattering is not only a more sensitive technique across a full 50–4000 cm^{-1} range of vibrations, but it has all the experimental advantages of spectroscopy in the visi-

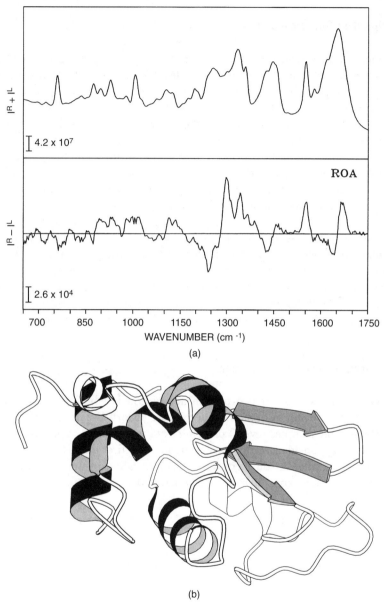

Fig. 4.38 (a) Raman circular intensity sum (upper trace) and difference (lower trace) spectra of (b) lysozyme in aqueous solution

ble/uv region. For example, by making use of resonance conditions, it is possible to study the stereochemistry of specific sites of biological activity in large biomolecules in aqueous solution. The same principles can be applied to optically inactive (achiral) molecules by inducing chirality with a magnetic field.

4.6
Multiphoton Spectroscopy

Multiphoton processes involve the concerted interaction of two or more photons with individual atoms or molecules. We have already dealt with some examples: simple Raman scattering is one case, in which one photon is absorbed and one is emitted in each molecular transition. Although the term 'multiphoton' is seldom used to describe this process, it does illustrate the essential concerted nature of the interaction, in that the absorption and emission are inextricably bound up with each other; it is not the same thing as absorption followed by fluorescence. However, the term 'multiphoton' is generally applied to processes involving the concerted absorption of two or more photons. The four-photon interactions connected with stimulated Raman scattering and CARS provide good examples, but the simplest multiphoton processes are those in which only absorption is involved. Figure 4.39 illustrates the essential difference between (a) a process involving two sequential single-photon absorption events, and (b) the concerted process of two-photon absorption. The former takes place only if both photons have suitable energies, and it results in population of both levels E_1 and E_2; in the latter case the only restriction is on the *sum* of the photon energies, and only level E_2 becomes populated.

4.6.1
Single-Beam Two-Photon Absorption

The very simplest case of two-photon absorption is where there is a single monochromatic beam of light and photons are absorbed pairwise by atoms or molecules of the sample. Several principles of more general application can be understood by considering this case in detail. The first thing to note is that it requires a very intense source of light for observation of the process; in fact, the effect was not experimentally demonstrated until the first pulsed lasers

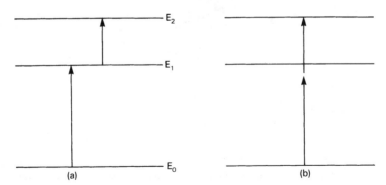

Fig. 4.39a and b Schematic energetics of (a) two sequential single-photon absorption processes and (b) two-photon absorption

arrived on the scene. It is not hard to understand why this is so. Because the photons have to be absorbed in a concerted fashion, it is necessary for two photons to pass essentially *simultaneously* through the region of space occupied by one molecule (or else the space occupied by a chromophore, in the case of localised absorption in a large molecule). The likelihood of this depends on the intensity of light and the molar volume and can be estimated quite easily.

For comparison purposes, we first note that for an unfocussed cw laser light source, for example an argon laser producing an irradiance of 10^7 W m^{-2} at a wavelength of 488.0 nm, the photon density given by Eq. (1.6) is approximately 7×10^{16} photons per m^3, corresponding to 2×10^{-12} photons per molecule in liquid water. Hence, even with this level of intensity, only one molecule in 5×10^{11} is experiencing the transit of a single photon at any instant of time. The probability of *two* photons simultaneously traversing a water molecule can be calculated from equation 1.30, with $M = 2 \times 10^{-12}$ and $N = 2$; the result is 2×10^{-24}, corresponding to only one in 5×10^{23}, in other words virtually one molecule per mole. We must conclude, then, that even if we had a molecule possessing the correct energy level spacing, there would be no likelihood of ever observing two-photon absorption with this level of intensity. However, if the same laser is mode-locked and focussed, peak intensities of 10^{15} W m^{-2} can easily be produced. Whilst this increase by a factor of 10^8 results in a roughly proportional increase in the probability of finding *one* photon in a molecular volume, the probability of finding *two* photons increases by a factor of approximately 10^{16} to roughly one molecule in 5×10^7. With *pulsed* lasers, then, we may expect to be able to detect two-photon absorption in suitable media.

Two other facets of two-photon absorption are illustrated by these considerations. Firstly, although it is not of much chemical interest, it is worth noting that if the calculations are repeated for thermally produced radiation of the same mean intensity, using Eq. (1.29) rather than (1.30), it turns out that the probability of finding two photons in a molecular volume is twice as large. This illustrates the fact that, unlike conventional absorption, multiphoton absorption is influenced by the photon statistics of the source. Much more important, however, is the fact that the probability, and hence the rate of two-photon absorption, depends *quadratically* on the irradiance. Thus, on passage of a beam of light through a medium exhibiting the effect, the rate of loss of intensity is given by

$$-dI/dl \propto I^2C, \tag{4.17}$$

in contrast to Eq. (4.1). The solution to this equation takes the form

$$I = I_0/(1 + \beta lC), \tag{4.18}$$

where β is a constant depending on the strength of the two-photon transition. Hence, the usual Beer-Lambert law of Eq. (4.2) does not hold for this process, and the concepts of absorbance and molar absorption coefficient do not apply as normally defined.

We now turn to the specifically spectroscopic properties of the two-photon process. Again, there is a substantial difference from conventional absorption in several areas. Foremost is the fact that the selection rules are quite different. As illustrated in Fig. 4.40, although the two-photon transition is subject to the condition

$$2h\nu = \Delta E, \tag{4.19}$$

on the laser frequency ν, it is not necessarily the case that the same transitions can be induced by single-photon absorption from a beam of frequency 2ν. In centrosymmetric molecules, for example, whilst the usual absorption selection rules permit only transitions between states of *opposite* parity, i.e. gerade-ungerade, the two-photon selection rules permit only transitions between states of the *same* parity, i.e. gerade-gerade or ungerade-ungerade. Indeed, two-photon absorption in the uv/visible range stands in the same relationship of complementarity to conventional uv/visible spectroscopy as Raman spectroscopy does to infra-red. Even for compounds with transitions which *are* both one- and two-photon allowed, two-photon spectroscopy in the visible range usefully provides access to states which would otherwise require uv excitation, as also shown in Fig. 4.40. In such cases, the ultraviolet response can thus be probed without the need for special vacuum techniques, as only *visible* wavelengths are involved.

Another interesting feature is the possibility of *Doppler-free spectroscopy*. If a mirror is placed such that the laser beam traverses the sample in both directions, then as shown in Fig. 4.41a and b, it is effectively irradiated by two counterpropagating beams of wave-vectors k and −k. The two Doppler-shifted

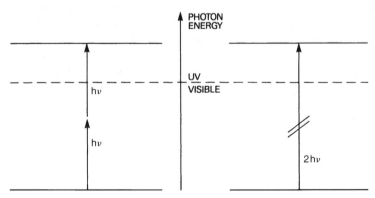

Fig. 4.40 Illustration of a transition which is two-photon allowed but single-photon forbidden

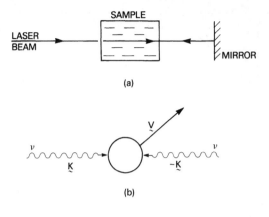

(a)

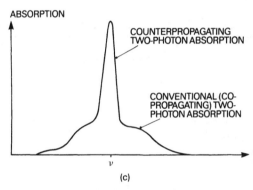

(b)

(c)

Fig. 4.41a - c Doppler-free two-photon spectroscopy

frequencies 'seen' by a sample molecule travelling with velocity v are given by the absorption analogue of Eq. (1.24), i.e.

$$\nu'_k \approx \nu(1 - v_k/c),\tag{4.20}$$

and hence equally

$$\nu'_{-k} \approx \nu(1 + v_k/c).\tag{4.21}$$

Two-photon transitions in which counterpropagating photons are absorbed thus provide an excitation energy in which the two Doppler shifts essentially cancel out:

$$\Delta E = h\nu'_k + h\nu'_{-k} = 2h\nu.\tag{4.22}$$

The result is a very sharp and symmetric Doppler-free absorption line on top of a broader band due to the two-photon absorption of photons travelling in the same direction, for which the usual Doppler shift occurs; a typical result is illustrated in Fig. 4.41c.

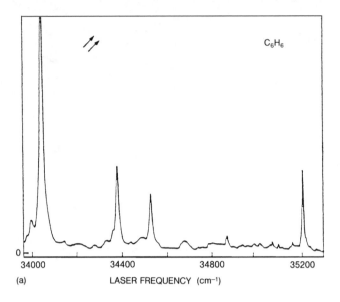

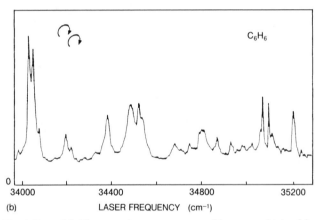

Fig. 4.42a and b The two-photon spectrum of benzene obtained by Whetten et al. (**a**) with plane polarised light, and (**b**) with circularly polarised light from a frequency-doubled dye laser. Reprinted from [17] by permission of the American Institute of Physics

The last major spectroscopic difference between two-photon absorption and convenional absorption is that even in isotropic samples such as gases or liquids, the two-photon process is very sensitive to the polarisation of the laser beam. In the conventional (single-photon) absorption spectroscopy of fluid samples, such effects are generally very small and are manifested only by optically active compounds. However, the whole appearance of a two-photon spectrum can be changed dramatically if the beam polarisation is altered, as illu-

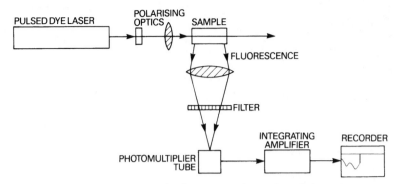

Fig. 4.43 Instrumentation for fluorescence-detected single-beam two-photon absorption studies

strated in Fig. 4.42. The precise way in which the intensity of each two-photon absorption band changes with laser polarisation provides useful information on the type of transition responsible. For example, in any two-photon transition from the usual totally symmetric ground state to a state which is not totally symmetric (in other words, one which lacks the full symmetry properties of the molecule), then the intensity of the associated absorption band increases by a factor of exactly $\frac{3}{2}$ on changing from plane to circular polarisation.

The recording of two-photon excitation spectra can most simply be accomplished either by directly monitoring the absorption from the laser beam or else by detecting the fluorescence due to subsequent relaxation of excited molecules in the sample. The former method is not easy, since the fractional absorption is generally very small indeed; it is simpler to detect the fluorescence which is relative to an essentially zero background. The apparatus for fluorescence-detected two-photon absorption spectroscopy is illustrated in Fig. 4.43 and is similar in principle to the apparatus used for the first demonstration of two-photon absorption by Kaiser and Garrett in 1961. A filter is used to cut out any fluorescence near to and below the irradiation frequency; any fluorescence observed at appreciably higher frequencies has to result from multiphoton excitation. An alternative detection method based on ionisation of the two-photon excited state is discussed in Sect. 4.6.3.

4.6.2
Double-Beam Two-Photon Absorption

At the molecular level, the types of transition involved in double-beam two-photon absorption are very similar to those described above. The energetics are such that

$$h\nu_1 + h\nu_2 = \Delta E, \tag{4.23}$$

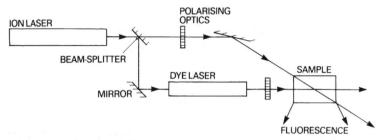

Fig. 4.44 Modification of the instrumentation in Fig. 4.43 for double-beam excitation. The angle between the two beams incident upon the sample will usually be smaller than shown here, to increase the intersection volume

and the rate of absorption is proportional to the *product* of the intensities of the two beams. In contrast to the single-beam method, however, the frequencies, directions and polarisations of the two absorbed photons can be independently varied. Typically, one beam could be derived directly from a primary laser source and the other from a dye laser cavity pumped by the same primary laser, with various polarising and reflective optics as shown in Fig. 4.44; the fluorescence is detected by a photomultiplier tube and the signal passed to a recorder as before. There are two main advantages to be had from use of double-beam methods.

The first advantage is the possibility of resonance enhancement, similar to that which we discussed in connection with the Raman effect (Sect. 4.5.1). If one of the beams has a frequency somewhere close to an optical absorption band of the sample, then the rate of two-photon excitation is greatly increased and a stronger signal is obtained. The second advantage is the facility to study the variation in the two-photon spectrum as both the polarisations of the two beams and the angle between them is varied. A very clever scheme devised by Martin McClain [18] involving measurement of the spectrum under three different sets of conditions provides the most complete information on the symmetry of the two-photon excited states and is a powerful tool for the determination of molecular electronic structure.

The types of substance in which two-photon absorption has been detected in the uv/visible range now extends from inert gases to human chromosomes. The new atomic and molecular excited states to which the effect provides access may have photochemical significance, although this is still a largely unexplored area. As a spectroscopic technique, two-photon absorption is very sensitive, providing the facility for single-atom detection in optimum cases. An intriguing technical application lies in the possibility of cutting three-dimensional patterns in clear plastic by focussing two laser beams inside it and moving the point of intersection around. Providing the frequencies are correctly chosen, two-photon absorption can only occur where the beams cross, initiating photochemical reaction to soften or harden the plastic. It has also been shown that much the same technique offers a basis for three-dimensional fluorescence imaging and, through induced photochromism, a novel kind of three-dimensional optical memory.

4.6.3
Multiphoton Absorption Spectroscopy

Multiphoton studies where more than two photons are absorbed are generally based on a single beam of laser light, and transitions are subject to the condition

$$mh\nu = \Delta E, \tag{4.24}$$

where m is an integer. Once again, the selection rules differ from conventional absorption, and indeed are distinctive for each value of m. The intensity of absorption, in the absence of resonance enhancement, depends on I^m, where I is the laser beam irradiance. Each additional photon essentially reduces the absorption signal by a factor of I/I_M, where I_M is the irradiance that would lead to ionisation or dissociation, and would therefore have a typical value in the region [19] of $10^{18\pm4}$ W m^{-2}. This figure certainly exceeds the level of irradiance in most laser-excitation experiments, and the rate of multiphoton absorption thus in general decreases very rapidly with any increase in the number of photons involved. In fact very few spectra involving more than three photons in the excitation process have been recorded, and these have generally required special techniques as discussed below. However, one distinction should be made before proceeding further. Whilst three and four-photon absorption generally represent the limit for uv/visible *spectroscopic* purposes, transitions involving the absorption of many more photons have certainly been observed in the *infra-red* laser excitation of polyatomic molecules. We shall return to this subject in Chap. 5; suffice it for the present to note that resonance enhancement is strongly involved.

The resonance aspect of multiphoton absorption lies at the heart of probably the most successful method for its detection, *multiphoton ionisation spectroscopy*. We saw in Sect. 4.2.2 that the detection of ions is a highly sensitive method in absorption spectrometry, and this sensitivity is especially well suited to the small signals expected from multiphoton studies. Since the method is applicable to gaseous samples, spectral resolution is also good. The principle involved is illustrated in Fig. 4.45. As the laser frequency is increased over its tuning range, the number of photons required to promote molecules from their ground state to the ionisation continuum drops from four to three, and the ion current increases accordingly. However, by far the largest signals are obtained under resonance conditions when the energy of one, two or three photons coincides with that of a bound state, as shown. For example at frequency ν_C, although three photons are responsible for the ionisation process itself, it is the two-photon resonance that results in a peak in the spectrum.

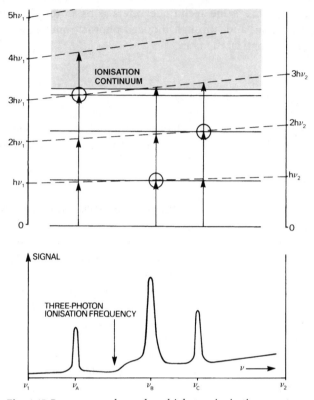

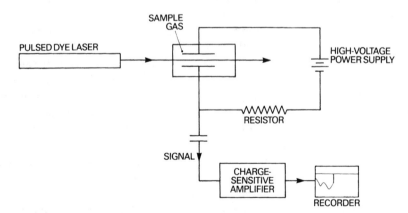

Fig. 4.45 Resonance-enhanced multiphoton ionisation spectroscopy. As the laser frequency is tuned from ν_1 to ν_2, multiphoton resonances occur at ν_A (three-photon), ν_B (one-photon) and ν_C (two-photon). With each resonance, the ionisation signal increases as shown in the lower half of the diagram

Fig. 4.46 Schematic apparatus for multiphoton ionisation spectroscopy

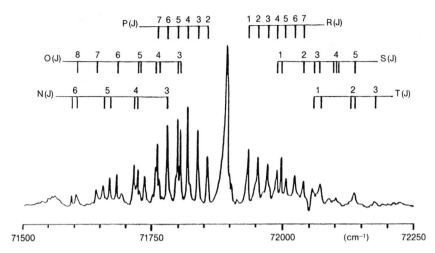

Fig. 4.47 Three-photon excitation spectrum of the H_2S $\tilde{H}^1B_1 \leftarrow \tilde{X}^1A_1$ transition obtained by resonant four-photon ionisation. Reprinted from [20] by permission of North Holland Physics Publishing

A typical apparatus used to obtain a multiphoton ionisation spectrum is illustrated in Fig. 4.46, and Fig. 4.47 shows part of the three-photon excitation spectrum of H_2S, obtained by Dixon et al. [20] using this method. The principle of multiphoton ionisation is also made use of in a laser mass spectrometer, which is discussed in Sect. 4.7.

4.6.4
Hyper-Raman Spectroscopy

The hyper-Raman effect is in many senses a hybrid between multiphoton absorption and Raman scattering. The process takes place in samples irradiated by an intense laser beam and involves three-photon transitions in which two photons are absorbed and one is emitted. Since two photons are absorbed from the laser beam in each transition, the hyper-Raman effect has a quadratic dependence on intensity and can only be detected using pulsed lasers; even with mode-locked laser pulse intensities in the region of 10^{15} W m^{-2}, the intensity of hyper-Raman scattering is approximately 10^{-6} times weaker than that of Raman scattering.

The molecules involved in hyper-Raman scattering may undergo transition either to a state of higher energy or to one of lower energy, as illustrated in Fig. 4.48: the two cases correspond to 'Stokes' and 'anti-Stokes' hyper-Raman scattering. Because of the weakness of the effect, it is usually only the relatively more intense Stokes transitions that are observed; these result in the production of frequencies ν' given by

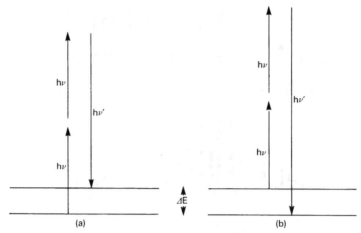

Fig. 4.48a and b The hyper-Raman effect: (a) Stokes transition, (b) anti-Stokes transition

$$h\nu' = 2h\nu - \Delta E. \tag{4.25}$$

Hence the energy uptake is measured in frequency terms by a shift away from the *harmonic* frequency 2ν. The selection rules for hyper-Raman scattering are very different from those which apply to conventional Raman scattering; for example, in centrosymmetric molecules, only *ungerade* vibrations show up in the spectrum. However, many more transitions are allowed than in single-photon absorption spectroscopy, and indeed one of the principal advantages of the hyper-Raman effect is its use for the direct observation of fundamental vibrations which are otherwise made manifest only in overtone or combination bands.

4.7
Laser Mass Spectrometry

The last topic to be discussed in this chapter is one which has brought about something of a revolution in the well-established field of mass spectrometry. This technique involves sample ionisation using a laser rather than the traditional electrical methods. Since ionisation of most molecules requires energies in excess of that which can be supplied by a single photon unless very short wavelengths below 150 nm are used, laser ionisation generally has to involve a multiphoton excitation process; hence, the term *multiphoton ionisation mass spectrometry* is also applied to this method. The advantage of using a mass spectrometer is that it enables the different types of ion produced by multiphoton absorption and subsequent fragmentation to be distinguished.

A typical laser mass spectrometer is based on a pulsed dye laser pumped by a harmonic of the Nd:YAG laser; by incorporating a frequency-doubling cap-

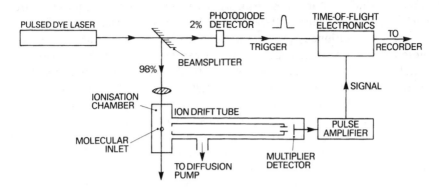

Fig. 4.49 Instrumentation for time-of-flight laser mass spectrometry. The molecular inlet is normal to the laser beam and is at its focus

ability such a system can resonantly ionise about half the chemical elements and nearly every kind of compound. Many details of the ionisation cell are determined by the nature of the sample. Samples are often introduced into the system in the gas phase, at vapour pressures usually not less than 10^{-6} atm but in some cases as high as 50 atm. Supersonic molecular beams are also amenable to laser ionisation analysis, and even liquid samples can be studied directly. Compared to a conventional instrument, the ionisation efficiency may be several orders of magnitude better, approaching 10% for the ground state molecules. However, the ionisation volume is much smaller, again by several orders of magnitude, in a laser setup. The ionisation volume can of course be increased by using an unfocussed laser, but this has to be played off against the resultant reduction in multiphoton ionisation efficiency. Detection methods also vary, as in other mass spectrometers, but one of the most effective is time-of-flight analysis, as illustrated in Fig. 4.49. Here, the various ion fragments are electrically accelerated and pass along a tube about a metre long. The time taken by each ion to reach the detector is determined by its mass and is typically measured in microseconds. Since laser pulses of only nanosecond duration are used to produce the ionisation, the mass spectrum can be obtained by monitoring the ion current as a function of the time elapsed since the laser pulse.

One of the most important aspects of laser mass spectrometry is the fact that the mass spectrum changes with the laser wavelength. As discussed earlier and illustrated in Fig. 4.45, the multiphoton ionisation process is most efficient when a resonance condition applies. A different mass fragmentation pattern can thus be recorded at each resonance, so providing a great deal more information on the molecular structure of the sample. Indeed, the mass spectrum can be represented by a three-dimensional plot, where ion current is plotted against mass in one direction and laser wavelength in the other. This technique is known as *resonance-enhanced multiphoton ionisation* (REMPI) mass spectrometry. Since multiphoton absorption generally produces molecular fragmentation patterns quite different from those of the normal mass spectrum, this method additionally provides a useful insight into the dynamics of multiphoton excitation. Whilst these methods can be used directly on chemically complex samples, it is also possible to treat samples by conventional 'wet chemistry' techniques to produce solutions for resonance ionisation mass spectrometry. This kind of approach is particularly expedient in analysis for inorganic elements for example Fe atoms can be detected by their resonant two-photon ionisation at a wavelength of 291 nm. By varying the irradiation wavelength, different elements can be selectively ionised and quantitatively measured in the mass spectrometer.

Other types of laser mass spectrometry are increasingly being directed to analytical applications. One such technique which has experienced spectacular growth is *matrix-assisted laser desorption ionisation* (MALDI). Here a small amount of sample is mixed with a suitable matrix material, commonly a low molecular weight organic substance such as glycerol, and subjected to a pulsed ultraviolet (usually nitrogen laser) beam which it strongly absorbs. Employing irradiances of between 10^{10} and 10^{12} W m^{-2} the sample *desorbs* as molecular or quasi-molecular ions, which are then analysed by conventional time-of-flight instrumentation. Such methods prove perfectly amenable to a wide range of large, involatile and thermally sensitive biomolecules such as proteins. Moreover with a mass range extending beyond 500 000 Da, highly accurate and rapid molecular weight measurements can be made; a complete mass spectrum can generally be obtained in a matter of microseconds. The technique is extremely sensitive, with often only femtomolar quantities consumed in the analysis procedure.

Higher levels of intensity in excess of 10^{12} W m^{-2} are employed for microsampling in *ablation* processes associated with the cavitation of solids. Mass spectrometric applications are here represented by *laser microprobe mass analysis* (LAMMA), or *laser-induced mass analysis* (LIMA). Based on the microprobe concept discussed earlier, such techniques are ideally suited to the point

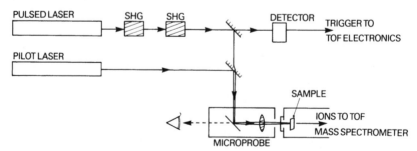

Fig. 4.50 Laser microprobe mass analysis apparatus for transmission sampling. The details of the time-of-flight mass spectrometer are as in Fig. 4.49

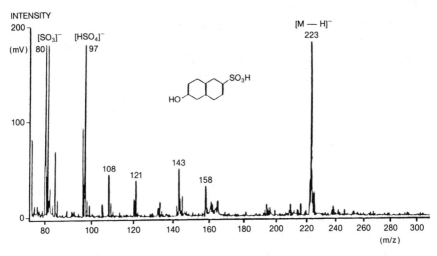

Fig. 4.51 Negative ion LAMMA spectrum of 2-hydroxynaphthalene-6-sulphonic acid, obtained by Holm et al. under normal atmospheric conditions. Reprinted with permission from [21]. Copyright 1984, American Chemical Society

analysis of surfaces. As shown in Fig. 4.50, high-power pulses of frequency-quadrupled light from a Q-switched Nd:YAG laser are focussed onto small areas of the sample, typically vapourising volumes of about 1 μm^3 per shot. A pilot He-Ne laser beam follows a collinear path onto the sample surface to facilitate visual location of the focus point. Ions released from the sample are analysed in a time-of-flight mass spectrometer. The method as illustrated is appropriate for transmission samples, but a reflection mode is commonly employed for characterising different areas of inhomogeneous surfaces or surface contaminants. The method can readily be applied to the subsurface analysis of inhomogeneous solids, successive laser pulses vapourising and releasing ions from progressively deeper layers within the sample. Such techniques have established extensive applications ranging from geology to microelectronics; they are particularly well suited to the study of fibres, environmental particles and metal corrosion processes, frequently out-performing conventional methods of microanalysis. The negative ion LAMMA spectrum of 2-hydroxy-naphthalene-6-sulphonic acid is shown in Fig. 4.51.

4.8
References

1. Pine AS, Maki AG, Robiette AG, Krohn BJ, Watson JKG, Urbanek T (1984) J. Am. Chem. Soc. 106:891
2. Lipson RH, LaRocque PE, Stoicheff BP (1985) J. Chem. Phys. 82:4470
3. Hurst GS (1987) Phil. Trans. Roy. Soc. Lond. A323:155

4. Snook RD, Lowe RD (1995) Analyst 120:2051
5. Collins CB (1985) J. de Phys. Colloq. C7:395
6. Saykally RJ, Veseth L, Evenson KM (1984) J. Chem. Phys. 80:2247
7. Mito A, Sakai J, Katayama M (1984) J. Mol. Spec. 103:26
8. Okuyama K, Hasegawa T, Mikami N (1984) J. Phys. Chem. 88:1711
9. Kachin SV, Smith BW, Winefordner JD (1985) Appl. Spec. 39:587
10. Demidov AA (1994) Biophys. J. 67:2184
11. Mohlmann GR (1985) Appl. Spec. 39:98
12. Hendra P, Mould H (1988) Int. Lab 18:34
13. Lutz M (1984) In: Clark RJH, Hester RE (eds) Advances in Infrared and Raman Spectroscopy, Wiley, Chichester, vol 11 p 211
14. Esherick P, Owyoung A (1982) In: Clark RJH, Hester RE (eds) Advances in Infrared and Raman Spectroscopy, Wiley, Chichester, vol 9 p 130
15. Hayashi S, Samejima M (1984) Surf. Sci. 137:442
16. Barron LD, Hecht L, Bell AF, Wilson G (1996), Appl. Spec. 50:619
17. Whetten RL, Grubb SG, Otis CE, Albrecht AC, Grant ER (1985) J. Chem. Phys. 82:1115
18. McClain WM (1971) J. Chem. Phys. 55:2789
19. Eberly JH, Maine P, Strickland D, Mourou G (1987) Laser Focus 23:84
20. Dixon RN, Bayley JM, Ashfold MNR (1984) Chem. Phys. 84:21
21. Holm R, Kämpf G, Kirchner D, Heinen HJ, Meier S (1984) Anal. Chem. 56:690

4.9
Questions

1. Describe briefly both the spectroscopic technique and the type of laser you would choose for each of the following problems, carefully explaining the reasons for your choice:
   (a) Mapping the surface concentration of a particular chemical component in a heterogeneous geological sample
   (b) Detection of trace quantities of lithium in aqueous solution
   (c) Determination of the stereochemistry of a chiral centre in a chromophore site of a large biomolecule
   (d) Study of the vibrations of copper phthalocyanine molecules adsorbed on a silver surface
   (e) Collection of a high-resolution infra-red spectrum of a gas-phase hydrocarbon
   (f) Direct detection of trace elements in blood
   (g) Identification of a transient radical species formed in the combustion of a hydrocarbon
   (h) Remote sensing of oil spills at sea

2. Compare and contrast the three fluorescence techniques termed *laser-excited fluorescence (excitation spectroscopy)*, *laser-induced atomic fluorescence*, and *laser-induced molecular fluorescence*.

3. Draw energy-level diagrams illustrating two possible configurations for optical-optical double resonance.

4. The Stokes Raman spectrum of ethene obtained using 488.0-nm radiation from an argon laser contains a line at 522.2 nm. Calculate the wavenumber of the vibration giving rise to this line and, using Eq. (4.11), calculate the wavelength and relative intensity of the corresponding anti-Stokes line for a spectrum recorded at 298 K. ($c = 3.00 \times 10^8$ m s^{-1}; $h = 6.63 \times 10^{-34}$ J s; $k = 1.38 \times 10^{-23}$ JK^{-1}).

5. Draw energy level diagrams illustrating the energetics of each of the following types of Raman scattering: (a) Stokes; (b) anti-Stokes; (c) pre-resonance Stokes; (d) resonance Stokes; (e) CARS; (f) inverse Raman Stokes; (g) inverse Raman anti-Stokes; (h) hyper-Raman Stokes.

6. Discuss the various means by which it is possible to prevent the interference of sample fluorescence in Raman spectroscopy.

7. Laser light often approximates to a Poisson distribution of photons, and the probability of finding N photons in a given volume can thus be calculated using equations (1.30) and (1.6) of Chap. 1. Calculate the probabilities of finding (a) one photon and (b) two photons within the volume occupied by a single molecule of a liquid with molar volume 8.07×10^{-5} m^3 mol^{-1}, irradiated by a mode-locked laser producing an irradiance of 10^{15} W m^{-2} at a wavelength of 444 nm. Repeat both calculations for an irradiance ten times as large. Comment on the implications of your results for two-photon absorption studies. ($c = 3.00 \times 10^8$ m s^{-1}; $h = 6.63 \times 10^{-34}$ J s; $L = 6.02 \times 10^{23}$ mol^{-1}).

8. Describe three of the principal features that distinguish two-photon absorption from single-photon absorption processes. Draw diagrams illustrating the time-varying intensity of a beam of laser light before and after passing through a medium which exhibits two-photon absorption.

9. A molecule moving in a laser beam of frequency ν experiences a relativistically shifted frequency ν' given by the formula $\nu' = \nu(1 - \beta)^{1/2}(1 + \beta)^{-1/2}$, where $\beta = v_k/c$ and v_k is the component of the molecular velocity in the direction of the beam. By taking a Taylor series expansion of this equation, show that for two-photon absorption from counterpropagating beams of the same frequency, the total energy absorbed is independent of the molecular velocity, if only terms linear in β are considered. Find the leading correction term for the energy and comment on whether or not it is likely to be significant for gas-phase two-photon spectroscopy.[5]

---

[5] The implication is a change in the molecule's kinetic energy associated with an increase in mass of E/c^2, where $E = 2h\nu$. Einstein obtained this result for the converse double-emission process in 1905.

CHAPTER 5

Laser-Induced Chemistry

> *May not bodies receive much of their activity from the particles*
> *of light which enter into their composition?*
> 'Opticks', Isaac Newton

In the preceding chapters, we have looked at a wide range of applications in which the laser is used as a probe for systems of chemical interest. Although the application of laser spectroscopic techniques in particular may result in short-lived changes in molecular energy level populations, the laser does not generally induce any *chemical* change in the sample; in that sense it is used as a static rather than a dynamic tool. Quite distinct from this is the field of applications in which laser excitation is used specifically to promote chemical reaction. This is a less well explored area and one that largely remains within the province of research and development, commercial applications being few and far between. Nonetheless, the subject encompasses a wide variety of topics, as we shall see. To introduce the subject, we begin with a general overview of the major principles.

5.1
Principles of Laser-Induced Chemistry

Chemistry induced by optical excitation is by definition *photochemistry,* and the whole of laser-induced chemistry can thus be regarded as one part of this much wider field. Although lasers can replace other light sources in any conventional photochemistry, there is a significant number of laser-induced reactions that are not practicable with conventional light sources. In this chapter the emphasis is firmly placed on these more distinctive applications specifically requiring laser stimulation. First, then, we need to identify the particular characteristics of laser light which mark out laser photochemistry as a separate identifiable discipline. The two most important qualities are undoubtedly the monochromaticity and high intensity of a laser source.

Laser monochromaticity naturally lends itself to applications requiring the selective excitation of particular sites within a heterogeneous system or of one specific chemical species in a mixture of reactants. The exploitation of this selectivity is particularly manifest in studies based on pulsed infra-red laser

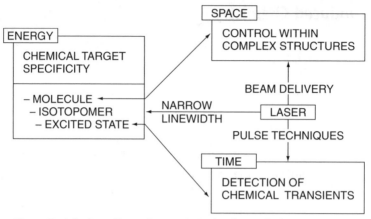

Fig. 5.1 Exploitation of laser characteristics for chemical selectivity in energy, space and time

radiation, which can offer a degree of reaction control quite unattainable by the traditional thermal methods of excitation. The generally high intensity of laser sources is significant both for increasing excitation efficiency and also for thereby promoting multiphoton processes; as we shall see, the latter feature is crucial for much of laser chemistry. Whilst the narrow linewidth of a typical laser source can be exploited in chemical selectivity based on molecular energetics, the usually tight collimation offers a spatial selectivity that can again be important in the photochemistry of structures such as surfaces, interfaces or other more complex systems. Here, facility to deliver the radiation through optical fibres can also prove useful, as, for example, in many medical applications. On the other side of the coin, the narrow beamwidth of a laser is, of course, a disadvantage for synthetic applications, in that without divergent optics and the associated loss of intensity, only very small volumes of material can normally be irradiated. Lastly, again in connection with the intensity, pulsed laser excitation offers a temporal selectivity that is now widely being exploited for inducing and monitoring fast and ultrafast chemical processes. A schematic representation of these key aspects of laser selectivity is shown in Fig. 5.1: we now look at the photochemical implications of these features in more detail.

5.1.1
Features of Laser Excitation

The obvious advantage of using an essentially monochromatic source of light for photochemistry is that only selected optical transitions are induced in the sample (although, of course, the subsequent photochemical reactions may involve states populated by decay processes rather than those directly accessed by optical absorption). This contrasts, for example, with the traditional use of a broadband flashlamp, which usually populates a number of different energy levels to an extent depending on the strength of each transition. These various excited states may lead to different types of chemistry and interfere with any particular photochemical reaction of interest. The same principle applies with even more force to chemical mixtures; here, a laser may be employed to selectively excite one specific component. This is especially useful in isotopically selective photochemistry, as we shall see in Sect. 5.4. The high intensity of monochromatic light available from lasers generally carries the advantage of being able to induce selective photochemistry with far greater efficiency than any filtered conventional source.

The powerful intensity of a laser is also significant in photochemical applications for another reason. Much photochemistry proceeds as a result of an initial excitation to electronically excited states, since it is electronic energies which are involved in the formation and rupture of chemical bonds. Whilst this excitation normally necessitates irradiation with the comparatively energetic photons of uv/visible light, we have seen at the end of the last chapter that by a process of multiphoton absorption it is possible for individual molecules to acquire more energy than is associated with a single photon. Hence, with a sufficiently intense laser source, even infra-red radiation can produce multiphoton excitation to electronically excited states and so lead to useful photochemistry. This represents a highly distinctive branch of photochemistry in which lasers are the only practicable sources. Again, there are important isotopically selective reactions which hinge on this type of laser photochemistry.

The ability to pulse laser radiation, quite apart from the high intensities which then ensue, provides the means for both inducing and monitoring ultrafast photochemical reactions. With the instrumentation described in Chap. 3, it is possible to identify short-lived transient species formed as intermediates during reaction. It is even possible to trace the course of processes which occur on a picosecond timescale, such as certain types of bond fission and the molecular rotations, torsional motions and electron transfer processes which play crucial roles in many biochemical reactions. Indeed, there are no other physical methods of measuring events which occur over such short times. An additional benefit accruing from the use of short pulses is that some of the slower processes which usually enter into the reaction kinetics, such as relaxation and diffusion, can effectively be suppressed; we shall pursue these subjects in more detail below.

Before considering the chemistry, brief mention should be made of one other general feature of laser excitation, and that is proper quantification of the absorption. Since the input radiation is commonly pulsed and so has time-variable intensity, and since both saturation and multiphoton absorption may further complicate the dynamics of photoabsorption, it is clearly no longer appropriate to employ the Beer-Lambert Law of Sect. 4.1. Instead, one is drawn to deployment of a simple descriptor of the net absorption such as the absorbance (optical density), defined through an obvious generalisation of Eq. (4.4) as

$$A = -\log_{10}(1 - F),\tag{5.1}$$

F being the fraction of energy absorbed. One can then gauge the often complex dependence on a multitude of factors such as: laser wavelength, fluence (energy density of the radiation), pulse duration, optical path length, temperature, and the concentration or pressure of both absorbing and non-absorbing species – a nice illustration is afforded by recent studies of the absorption of CO_2 laser radiation by methylphosphine [1]. Many of these factors thereby exert a significant effect on the rates and yields of the reactions that occur subsequent to absorption, as will be discussed in particular connection with multiphoton excitation in Sect. 5.2.3.

5.1.2
Laser-Initiated Processes

Let us first consider the various types of chemical process which can be initiated by the absorption of laser radiation. In polyatomic molecules, the initial photo-induced transition to an electronically excited state is almost invariably followed by some degree of intramolecular relaxation before any real chemistry takes place. Such unimolecular relaxation processes generally involve redistribution of energy amongst vibrational states and take place typically over nanosecond or sub-nanosecond timescales, as will be discussed in Sect. 5.3.4. The state directly populated by photon absorption may therefore have little *chemical* significance. Relaxation may lead to ionisation, isomerisation or dissociation, as illustrated for a polyatomic molecule ABC in the following scheme:

$$\text{Photoabsorption: } ABC + nh\nu \rightarrow ABC^* \quad (n \geq 1)\tag{5.2}$$

$$\text{Autoionisation: } ABC^* \rightarrow ABC^+ + e\tag{5.3}$$

$$\text{Isomerisation: } ABC^* \rightarrow ACB^*\tag{5.4}$$

$$\text{Dissociation: } ABC^* \rightarrow AB^\ddagger + C^\ddagger\tag{5.5}$$

The last of these processes results in bond fission and produces the fragments AB and C, which are generally both in vibrationally excited states, and in some cases are also electronically excited. The term 'photolysis', also occasionally applied to photoisomerisation, is most often applied to this kind of process in which dissociation follows absorption, in other words where molecular fragmentation is induced by the absorption of light. As such, it represents the simplest kind of photochemical reaction, and it also provides the mechanism for the first step in many more complex types of photochemistry. For example, if AB and C are radical species, laser photolysis can provide the initiation step for a chain reaction. The simplicity of Eq. (5.5) belies the complexity of many reactions involving laser-induced dissociation. In particular, many polyfunctional compounds can exhibit more than one type of dissociation, leading to a mixture of products. We shall return to a detailed consideration of specific unimolecular processes in Sect. 5.3.1.

Whilst the above considerations apply to unimolecular reactions, lasers can also be used to induce bimolecular reactions in which either one or both of the reactants are initially excited by the absorption of laser light. Even if photoabsorption only takes place in one of the reactants before the collisional reaction, the laser excitation can still play a crucial role in *state preparation*, since the chemistry of excited molecules often differs from their ground-state counterparts. In principle, a wide range of reaction conditions can be obtained by promoting each reactant to various energy levels. Often, infra-red lasers are used for this purpose, promoting reactants to relatively low-lying vibrationally excited states. The problem with electronic excitation is that after the absorption stage, processes (5.3) to (5.5) may occur in either reactant before there is time for any reactive collision. Of course, further laser excitation or fragmentation may also be involved at each intermediate stage. For these reasons the kinetics of laser-induced bimolecular reactions are frequently highly complex, as we shall discover in looking at some specific examples in Sect. 5.3.2.

Before leaving these general considerations, it is worth noting that there is one other intriguing possibility for laser-induced bimolecular reaction which does not fit into the scheme outlined above [2]. This is based on the fact that the products of bimolecular reactions are usually formed from reactants X and Y through a transient activated complex XY^*, which is energetically unstable. Here the reaction pathway through from reactants to products may be associated with a vibration of the complex; for example the reaction $H + D_2 \rightarrow HD + D$ essentially proceeds via an antisymmetric stretch of $H \cdot \cdot D \cdot \cdot D$. In such a case, laser irradiation at the appropriate frequency of the complex can enhance the reaction rate. This is a significantly different approach from most other laser-induced chemistry because the irradiation frequency does not correspond to an absorption band of the starting material; indeed, in the example given, the reactants are themselves infra-red inactive. However, relatively few studies appear to have been based on this notion.

5.2
Multiphoton Infra-red Excitation

A very distinctive kind of laser photochemistry can be induced by powerful infra-red sources, the carbon dioxide laser being by far the most widely used. As described in Chap. 4, the multiphoton processes which can be induced by intense radiation become particularly efficient if one or more resonance condition can be satisfied by the molecular energy levels. Where a single frequency of uv/visible radiation is involved, the unequal spacing of most *electronic* levels means that it is rare to obtain even one intermediate state resonance.

However, *vibrational* energy levels are more or less equally spaced, at least for the lowest levels of excitation. Hence, with infra-red radiation of the appropriate wavelength, multiphoton absorption can become highly significant.

5.2.1
Diatomic Molecules

To consider multiphoton infra-red absorption in more detail, we first take the simple case of a diatomic molecule, where there is only one vibrational frequency. The appropriate energy levels are shown in Fig. 5.2, together with arrows representing the absorption of infra-red photons with the same frequency. The first thing to note is that as we move up the ladder of vibrational states, although the spacing between adjacent levels starts off fairly constant, it diminishes at an increasing rate. Of course, it also has to be borne in mind that each vibrational level has its own manifold of much more closely spaced rotational levels. Because each of these levels has an associated linewidth, there comes a point at which we effectively have a quasi-continuum of states, as represented by the shaded area in the diagram. Eventually, an asymptotic limit is reached, at which point there is no longer any restoring force as the two atoms move apart, and dissociation occurs. The question of linewidth is one of the crucial considerations in multiphoton infra-red absorption, both in connection with the radiation and the molecular transitions. As far as the radiation is concerned, linewidths vary according to the nature of the source. The emission from a diode laser can have a linewidth as small as 100 kHz (3×10^{-6} cm^{-1}); at the other extreme the linewidth of a high-pressure CO_2 laser may be anything up to 10^6 times larger. Whilst in energy terms this is still very much smaller than the gap between the lowest-lying vibrational levels, this is not the case for vibrational levels approaching the dissociation limit. Regarding the linewidths of molecular transitions, the usual line-broadening mechanisms apply to an extent which largely depends on the state of the sample. However, power-broadening represents an additional factor which is especially important under conditions of high intensity irradiation. With laser irradiances of 10^{10}–10^{12} W m^{-2}, power-broadened linewidths typically lie in the 1–10 cm^{-1} region. Again, this contributes to the creation of a quasi-continuum near the dissociation limit.

The process of multiphoton absorption displays different characteristics over different regions of the energy scale, and it has become common to speak in terms of regions I, II and III, illustrated in Fig. 5.2. In region I, vibrational levels are quite widely spaced, and the spacing is greater than the overall absorption bandwidth. Because the spacing is non-uniform, however, the photon energy soon gets out of step, and multiphoton processes occur. In the diagram, for example, the transitions v = 0 → 1, 1 → 2, 2 → 3, 3 → 4 and 4 → 5 all require energies close to that of a single photon and lying within the overall bandwidth. These transitions therefore all take place by the process of single-

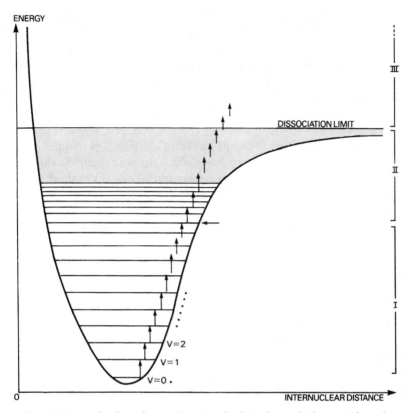

Fig. 5.2 Energy levels and transitions involved in the multiphoton infra-red dissociation of a diatomic molecule. Note that each vibrational level has rotational fine structure (not shown), and also the energies are not discrete but have finite linewidths; both these features are important in the excitation process

photon absorption. The energy required for the $5 \to 6$ transition, however, is sufficiently different that it lies outside the bandwidth and cannot take place by absorption of one photon. Nonetheless, excitation can proceed up to the $v = 10$ level, as indicated, by a direct $5 \to 10$ transition involving four-photon absorption. It is the main characteristic of region I that such concerted multiphoton processes take place on the way up the vibrational ladder. As we saw in Sect. 4.6, this necessitates a fairly intense flux of photons and, hence, a powerful source of radiation.

Region II is characterised by quasi-continuum behaviour resulting from the fact that vibrational energy level spacing has become *less than* the bandwidth. Here successive photons can be absorbed in a series of energetically allowed single-photon transitions. Since energy conservation is satisfied at every step, the molecule can at each point exist for a finite lifetime before absorbing the next photon; hence, excitation through this region does not necessitate the enormously large photon flux which might at first appear necessary. Finally, once the level of excitation has reached the dissociation threshold, a true energy level continuum is encountered, and further photons can be absorbed in the short time before the atoms separate; this is known as region III behaviour. Note that in sharp contrast to region I, the process of excitation through regions II and III is relatively insensitive to the precise irradiation frequency. The entire process leading to dissociation may involve the absorption of 30–40 infra-red photons.

In passing, we note that the term 'coherent excitation' is frequently applied to the multiphoton absorption processes which occur in region I. This refers to the fact that the interval between successive transitions is too short for relaxation to occur. This represents a distinction from the term 'incoherent excitation' applied to the sequential single-photon absorption processes occurring in region II. These are perhaps unfortunate expressions in that 'coherent' is already a rather over-used word with other very different meanings in connection with lasers. Quite apart from signifying the special phase properties of laser light discussed in Sect. 1.6.3, it has a very specific meaning in connection with multiphoton processes in general, as discussed in Sect. 4.5.2. Neither of these senses applies to either region I or region II multiphoton absorption; such processes can certainly take place with incoherent thermal light of sufficient intensity, and they are in the sense of wave-vector matching explicitly *incoherent* by nature. It is therefore preferable to adopt the term *concerted excitation* for non-resonant multi-photon absorption such as takes place in region I and *sequential excitation* for a series of single-photon absorptions as in region II.

5.2.2
Polyatomic Molecules

When we turn to the case of *polyatomic* compounds, many of the same principles apply, although as we shall see there are so many additional factors to consider that the detailed theory becomes very complex indeed. In polyatomic molecules containing N atoms, there are generally (3N–6) modes of vibration, each with its own ladder of vibrational and associated rotation–vibration levels. It is therefore possible to induce multiphoton excitation at a number of different wavelengths, corresponding to each of these various vibrational modes. There may also be several different processes of dissociation, involving the fission of different chemical bonds and, hence, leading to different types of fragmentation. Moreover, further multiphoton excitation may occur in any fragment with suitable vibrational levels, in some sense representing a fourth stage in the overall process of multiphoton decomposition. Nonetheless, in many cases it is not necessary to reach the dissociation level in the parent molecule before useful photochemistry can occur.

One of the most important and distinctive features of multiphoton infra-red absorption by polyatomics concerns the nature of the excitation in the quasi-continuum region II. The multiplicity of energy levels means that even in the gas phase, power-broadening results in formation of an effective continuum far below the dissociation limit, and region II behaviour thus commences at quite low energies. In comparatively small polyatomics, region I may span only a few vibrational quanta; in larger molecules it may be that only one quantum is absorbed before region II behaviour applies. For example in the case of SF_6, region II appears only a few thousand cm^{-1} above the zero-point energy; for molecules with twenty or more atoms it may appear as low as 500 cm^{-1}. In the liquid or solid state, inhomogeneous line-broadening also comes into effect (see Sect. 1.5.3) and contributes to quasi-continuum formation. Hence, the onset of continuum behaviour generally takes place much lower down the energy scale than is the case for a diatomic molecule. Another complication is the fact that many polyatomic molecules have electronically excited states lying below the dissociation limit of the electronic ground state. High levels of excitation thus result in some population of these electronic excited states and may, for example, lead to fluorescence decay (as notably in OsO_4) or further vibrational excitation.

Because there are a number of different vibrational modes in polyatomic molecules, there is a considerable amount of overlapping amongst the various energy levels in the quasi-continuum. Since the bandwidth of these levels is generally large compared to the spacing between adjacent levels, this results in a relative inselectivity over the vibrational mode into which the energy of each absorbed photon passes. Also, vibrational energy may be redistributed into different modes or combinations of modes as a result of anharmonic interactions; the time taken for this relaxation process is typically 10^{-11} s. Collision-induced intermolecular energy transfer also takes place, but over a longer timescale determined by the pressure (typically 10^{-8} s at 1 atm).

It is now generally recognised that under intense infra-red irradiation, the vibrations of the nuclear framework of a polyatomic molecule are in fact most realistically represented by an essentially random, or *stochastic*, mixture of the normal modes of vibration, between which there is completely free energy flow. The result has important implications for the energetics of the multiphoton excitation process. For example, in a certain molecule, absorption at one particular frequency (corresponding to a fundamental vibration) may in principle require only 30 photons to accomplish dissociation. Nevertheless, absorption of 30 such photons will in practice result in various amounts of energy being deposited in each of the vibrational modes, and the level of excitation will therefore fall far short of the dissociation limit. Consequently, it is commonly found that the mean number of photons absorbed per molecule far exceeds the energetic minimum and can in some instances run into hundreds.

Much more significant, however, is the fact that these vibrational relaxation processes greatly diminish the prospects for selectively populating *specific* vibrational modes. In many cases, the results of powerfully pumping a molecule with infra-red radiation can be virtually unchanged even when frequencies corresponding to two completely different bond stretches are employed; it is still *generally* the case that the weakest bond is the first to break. For example, the multiphoton dissociation of SF_5Cl produces SF_5 and Cl radicals even if the laser is tuned to the frequency of a sulphur–fluorine vibration. Thus, the very appealing notion of mode-selective chemistry, in which a laser could be used to excite and dissociate any chosen bond in a polyatomic molecule according to the exact wavelength employed, is not anywhere near as realistic as was at first hoped. It was, indeed, the expectation behind this largely unfulfilled promise that led to much of the initial funding of research in this area.

With *ultrashort* laser pulses, mode-selective excitation should be more feasible since vibrational relaxation would not have time to occur to any large extent, and the conditions *immediately* subsequent to each pulse might be very far removed from thermal equilibrium; however, it has only comparatively recently become possible to achieve the optimal picosecond pulse lengths in the infra-red. Despite these problems, there are nonetheless some laser-induced reactions that do take place before complete randomisation of vibrational energy occurs, and these can have unique and highly useful applications, as will become evident during the course of this chapter.

5.2.3
Reaction Rates and Yields

We have seen that mode-selective chemistry is seldom possible due to the stochastic nature of multiphoton vibrational excitation in polyatomic molecules. This is exactly in accordance with the highly successful *RRKM (Rice-Ramsperger-Kassel-Marcus) theory* of reaction kinetics, in which it is assumed that energy can flow completely freely between different vibrational modes, a process known as *intramolecular vibrational (energy) randomisation*. The energy efficiency of a laser-induced chemical process can thus be rather poor (although no poorer than the corresponding thermally induced process) since generally more photons are required to bring about reaction than would be expected from the activation energy. There are exceptions to this rule, however, as we shall see. It is often possible to exercise considerable control over the course of reactions induced by multiphoton excitation, and the interplay between kinetic and thermodynamic factors is seldom as clearcut as in other branches of chemistry. The major experimentally controllable factors listed below influence not only reaction rate and yield but also the relative yields in cases in which more than one product can be formed. In many, but not all, respects, the intricate nature of the dependence on these experimental variables reflects the character of the initial photoabsorption process discussed in Sect. 5.1.1.

1. Laser wavelength. As discussed above, whilst the monochromaticity of laser radiation in principle offers the possibility of delivering energy to specific

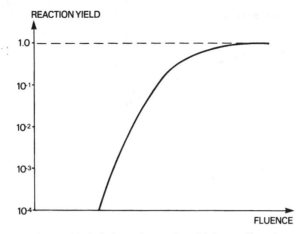

Fig. 5.3 Typical dependence of multiphoton dissociation on laser fluence, showing the onset of saturation behaviour

bonds or vibrational modes in a polyatomic molecule, relaxation processes rapidly redistribute this energy and so greatly diminish the localisation or selectivity of excitation. Nonetheless, even where substantial relaxation takes place before reaction, there is always some dependence on the wavelength of radiation applied to the system.

2. Fluence (energy density). Most laser chemistry is carried out with pulsed, rather than continuous-wave (cw) lasers, and the number of pulses applied obviously affects the extent of reaction. Although the sample irradiance during each pulse is determined by fluence, pulse duration and temporal profile, there is good evidence that, in the case of multiphoton excitation, it is fluence which exerts the major influence on the course of reaction, particularly in region II. A typical case is represented by Fig. 5.3, which shows the yield of a single-pathway reaction as a function of fluence. The yield generally increases with fluence, until saturation takes place at high energy densities. Not all reactions display the same kind of dependence, however. Other parameters such as pulse duration and repetition frequency generally influence the extent of excitation in a fairly obvious way, but are not always amenable to variation.

3. Pressure. Most laser-induced reactions take place in the gas phase at low pressure. The effect of increasing pressure, whilst often highly significant, is difficult to generalise since it influences the reaction in a number of different ways. The simplest effect is, of course, an increase in the number of molecules in the laser beam. The effect of pressure-broadening can also change the rate of absorption by increasing the absorption linewidth. Often more important, however, is the increase of collision-induced relaxation with pressure. The resultant increase in the rate of exchange of vibrational energy between excited

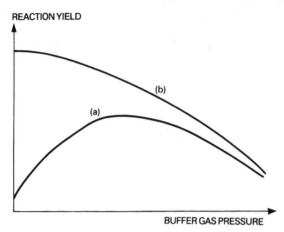

Fig. 5.4 Variation of multiphoton dissociation yield with
pressure of buffer gas; (**a**) where there is relatively little direct laser excitation of the reactant into the quasicontinu-
um region, and (**b**) where most reactant molecules are directly excited by the laser

and unexcited molecules can play a crucial role in populating the quasi-continuum leading to reaction. For example, the dissociation yield of silane, SiH_4,
increases very appreciably with pressure. In some reactions, however, such as
the dissociation of SF_6, the yield is relatively insensitive to pressure. The mean
number of photons absorbed by each molecule can also vary markedly with
pressure. This is because as pressure increases, collisions can supply the extra
energy required to reach the dissociation threshold.

4. Buffer gases. Lastly, there is the effect of admixing monatomic or relatively
simple polyatomic buffer gases such as argon or nitrogen, usually in considerable excess, with the reactant. If relatively few reactant molecules undergo direct laser excitation, transfer of vibrational energy by collision with buffer molecules can increase the extent of reaction in much the same way as an increase
in pressure. However, as the buffer gas pressure increases, deactivation of the
reactant will ultimately reduce the mean level of excitation and so decrease the
reaction yield, as shown in Fig. 5.4a. If the majority of sample molecules are
directly excited into the quasi-continuum by the laser, buffer gases can only
lead to a decrease in yield, as shown in (b).

A classic piece of work by Zitter and Koster [3] illustrates just how significant
some of these factors can be. In a study on laser-induced elimination of HCl from
CH_3CF_2Cl it was found that the mean number of photons absorbed per molecular dissociation decreased dramatically both with increasing pressure and cw
laser power. Using 966 cm^{-1} CO_2 laser radiation, the number of photons required was shown to vary between almost 500 at 100 torr pressure and 6 W laser
power to just over 5 photons at 600 torr pressure and 28 W power.

Many of the conventional kinetic and thermodynamic rules governing chemical reactions cannot be directly applied to laser-induced chemistry. A significant result of the study by Zitter and Koster [3] was the observation that under suitable conditions the mean energy requirement can be much *less* than the activation energy (in this case equivalent to 22 photons) and approach the value of the enthalpy change ΔH (equivalent to 5.2 photons) – this is possible because energy released as the products are formed from the activated complex can contribute towards the activation of other reactant molecules. However, the concept of enthalpy change at a given temperature is rather ill-defined for a laser-induced reaction since the entire system departs very markedly from thermodynamic equilibrium. It is principally for this reason that equilibrium constants can no longer be directly calculated from the usual relationship $K = \exp(-\Delta G/RT)$.

It is rare to apply heat in the case of a laser-induced reaction, since it results in non-specific thermal excitation of vibrational levels and thus reduces the prospects of observing any novel chemistry. In fact, it is a distinctive feature of most laser-induced reactions that they take place 'in the cold'. In the case of reactions which can be induced either by heat or by infra-red laser radiation, it is generally found that side-reactions take place to a much lesser extent where the laser is employed. This is largely a reflection of the fact that the reaction vessel does not have the hot walls at which reactive intermediates normally form.

5.3
Laser Photochemical Processes

Having discussed the basic principles of laser-induced chemistry, we now move on to a consideration of the various types of reaction that can occur. Although some examples will be given of reactions stimulated by uv/visible radiation, there are many other photochemical reactions that can, in principle, be laser-induced but which can equally be induced by other less intense, incoherent sources of light. Most of the more novel chemistry induced by lasers involves infra-red multiphoton excitation as described in the last section, and the following account accordingly places most emphasis on this area.

5.3.1
Unimolecular Laser-Induced Reactions

By far the largest number of laser-induced chemical reactions fall into the category of unimolecular reactions, and the carbon dioxide laser, producing powerful emission at numerous discrete wavelengths around 9.6 μm and 10.6 μm (see Table 2.1, p. 41), is the most commonly applied source. The simplest type of unimolecular reaction is isomerisation, and several studies have shown how laser-induced *photoisomerisation* can modify the relative proportions of different isomers in a mixture. Before considering examples of this laser-induced process, however, it is first worth noting the factors that normally determine these relative proportions.

It is often the case that an organic synthesis produces more than one isomer, and unless the reaction is kinetically controlled, the yield of each is determined by the position of equilibrium for the interconversion reaction. As mentioned earlier, the corresponding equilibrium constant is, in turn, usually related to the temperature and the relative thermodynamic stability of the isomers through the well-known relation $K = \exp(-\Delta G/RT)$. Except in cases where the value of ΔG is exceptionally small, the result is that only at very high temperatures can the less stable isomer be present in more than minute proportions. At such high temperatures, decomposition would in any case generally occur.

However, the selective laser excitation of one isomer, using a wavelength which no other isomer appreciably absorbs, can substantially modify the relative proportions either towards or, indeed, in some cases away from equilibrium. An example of the former case is afforded by 1,2-dichloroethene, where the cis-isomer is more stable than the trans-isomer by approximately 2 kJ mol^{-1}. Pulsed irradiation of a mixture containing an excess of the trans-compound at a frequency of 980.9 cm^{-1} results in conversion to a mixture in which the cis-isomer predominates [4]. Pulsed irradiation of hexafluorocyclobutene at 949.5 cm^{-1}, however, results in up to 60% conversion to its isomer hexafluoro-1,3-butadiene [5], which is thermodynamically less stable by 50 kJ mol^{-1}:

$$\rightarrow CF_2{=}CF{-}CF{=}CF_2 \tag{5.6}$$

This is simply a reflection of the fact that the cyclic compound absorbs radiation of this frequency much more strongly than its isomer. It is interesting to note that with sufficiently high intensities, further laser-induced reactions take place and lead to the formation of decomposition products and low-molecular-weight polymers [6].

A classic case of laser-induced chemistry involves the conversion of 7-dehydrocholesterol (I) (Fig. 5.5) to previtamin D$_3$ (II), which is, once again, an isomerisation reaction. The usual photolytic method by which this conversion is accomplished involves a series of fractional distillations and produces relatively low yield. However, by using two-step laser photolysis with KrF (248 nm) and N$_2$ (337 nm) laser radiation, it was shown that the competing side-reactions can largely be eliminated and the conversion effected with 90% yield [7]. Since the product (II) is reversibly convertible to vitamin D$_3$ (III), the overall process represented a useful and substantially improved method for synthesis of the vitamin. Much other recent work on ultraviolet laser chemistry concerns the photodissociation of relatively small molecules, a technique affording powerful insight into the detailed energies and angular properties of molecular electronic states.

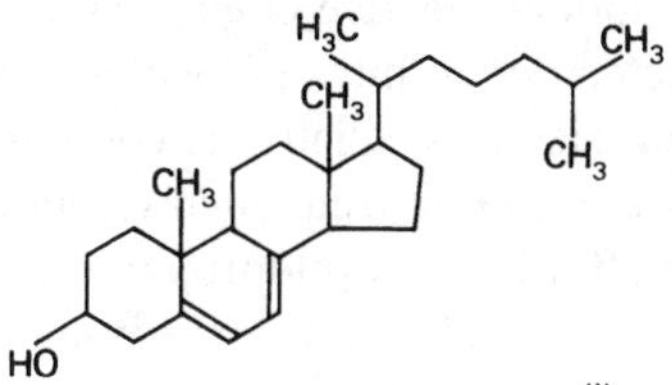

Fig. 5.5 Molecular structures of (I) 7-dehydro-cholesterol; (II) previtamin D$_3$; (III) vitamin D$_3$

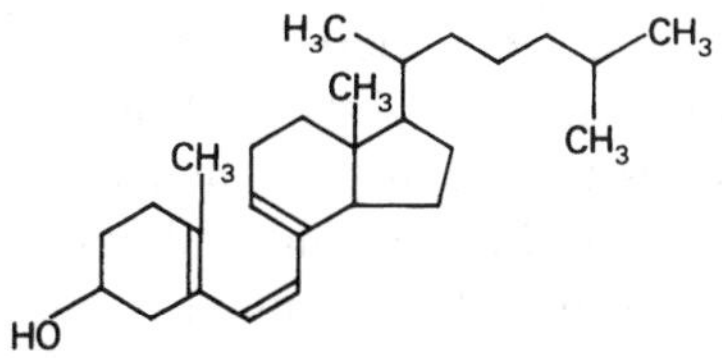

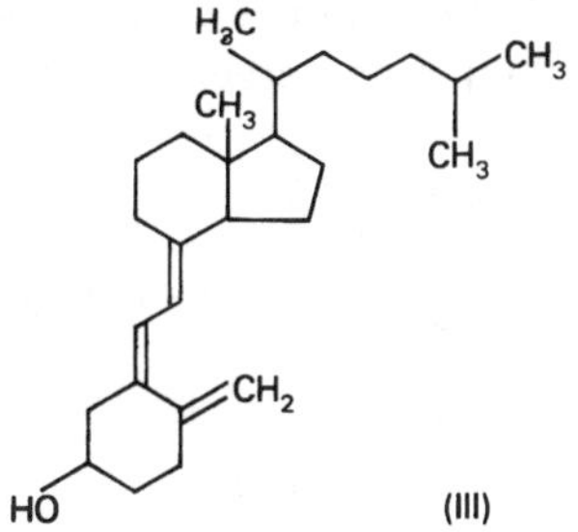

Most unimolecular laser-induced reactions involve multiphoton infra-red *dissociation*, with many organic elimination reactions coming under this heading. These reactions can often produce high yield since the molecular products are formed directly rather than through secondary reactions of radicals or other reactive intermediates. Some good examples are provided by elimination reactions involving esters, which proceed as follows:

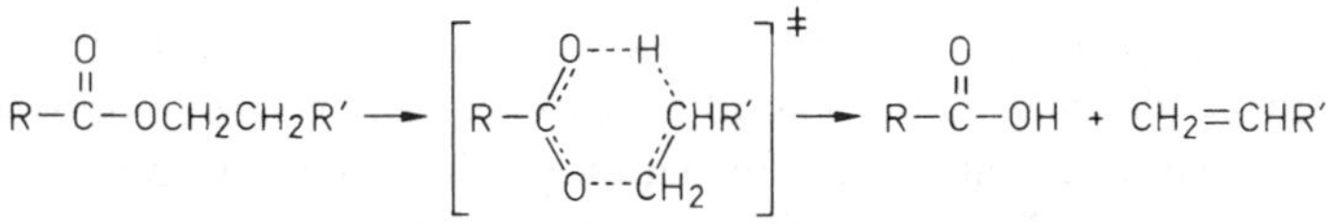

$$R-\overset{O}{\overset{\|}{C}}-OCH_2CH_2R' \longrightarrow \left[R-\overset{O---H}{\underset{O---CH_2}{C}}CHR' \right]^{\ddagger} \longrightarrow R-\overset{O}{\overset{\|}{C}}-OH \; + \; CH_2{=}CHR' \tag{5.7}$$

Such reactions can be very effectively induced by laser irradiation at a frequency of around 1050 cm^{-1}, which produces excitation of the stretching mode of the O–CH$_2$ bond and ultimately results in its fission.

As mentioned above, mode-selective chemistry is not generally as feasible as might at first be expected due to intramolecular relaxation processes. Nonetheless, there are certain cases, especially in comparatively small molecules, where irradiation at different laser frequencies genuinely results in different products. An example is afforded by cyclopropane, where it is found that multiphoton excitation at around 3000 cm^{-1} corresponding to the CH stretch frequency results in isomerisation to propene. However, irradiation at around 1000 cm^{-1}, corresponding to the CH_2 wag, produces both isomerisation and fragmentation in roughly equal amounts.

Other reactions which have been induced by multiphoton dissociation are far too numerous to mention. Often there is more than one end-product, and the product ratios are frequently very different from those obtained in the corresponding thermal pyrolysis experiments. The fact that more than one fragmentation product appears in the laser-induced reactions is, of course, clear evidence that energy is well distributed amongst various vibrational modes. In such cases the activation energy generally appears to be the major factor in determining the product ratios, although laser pulse power and energy can also exert a remarkably large influence. For example, in the following reactions [8], (5.8) overwhelmingly predominates when using laser pulses of 10^4 J m^{-2}, but with pulse fluences four times larger, (5.9) becomes the principal reaction:

$$\tag{5.8}$$

$$\tag{5.9}$$

Whilst the precise explanation for this not uncommon type of behaviour is still the subject of debate, it is likely that the competition between alternative relaxation channels with differing kinetics is responsible.

5.3.2
Bimolecular Laser-Enhanced Reactions

When laser radiation induces a *unimolecular* reaction, the products may of course participate in further chemistry by a *bimolecular* reaction with another reagent. This principle can be put to good effect, especially where unimolecular photodecomposition results in the formation of highly reactive radicals or other short-lived species. Quite apart from this type of process, however, there exists the possibility of exciting one reactant, without chemical change, to a state in which it is more reactive towards the other reagent. Mostly this is accomplished by irradiation at an infra-red frequency corresponding to a molecular vibration of one of the reactants, and the method is thus known as *vibrationally enhanced reaction*. Several attempts to induce such reactions have failed, most likely because for these reactions there is insufficient translational energy for reactive collisions without the application of heat (one way around this problem is to use molecular beams; see Sect. 5.3.5). However, there is also a large catalogue of successes, and in optimum cases the vibrational excitation of even a single quantum in one of the reactants can lead to an increase in the reaction rate constant by several powers of ten. Most of the successfully demonstrated reactions of this type involve the collisional transfer of single atoms between comparatively small molecules.

As a first example, we can take the reaction between ozone and nitric oxide:

$$O_3 + NO \rightarrow NO_2 + O_2 \tag{5.10}$$

The product nitrogen dioxide can be formed in either its electronic ground state ($^2A_1$) or an electronically excited state ($^2B_2$), and separate rate constants, k_1 and k_2 respectively, are associated with the corresponding rates of formation. Since the reaction involves fission of one of the O_3 bonds, the best way of enhancing the reaction by vibrational excitation is to pump the antisymmetric stretch. It so happens that the corresponding frequency coincides with the P(30) line (1037.4 cm^{-1}) in the 9.6 μm emission of the carbon dioxide laser, and experiment has shown that when this radiation is applied to the system, the reaction rates at 300 K increase by a factor of 4 for k_1, and 16 for k_2 [9].

Another example is the radical reaction between methyl fluoride and bromine:

$$CH_3F + Br \rightarrow CH_2F + HBr \tag{5.11}$$
$$CH_2F + Br_2 \rightarrow CH_2FBr + Br \tag{5.12}$$

Once again, vibrational excitation of the methyl fluoride can be accomplished with CO_2 laser 9.6 μm radiation, in this case the P(20) line at 1046.9 cm^{-1}. The result is an increase in the reaction rate by a factor of approximately 30 at 300 K [10].

Even neater experiments are possible if one of the reactants is a molecule which is itself the active medium in a molecular laser. This obviates the problem of searching for accidental coincidences between the emission lines of a laser source and the absorption features of the reactants. For example, the rate of reaction at 300 K between free (^3P) oxygen atoms and hydrogen chloride

$$O + HCl \rightarrow OH + Cl \tag{5.13}$$

can be enhanced by a factor of approximately 100 by irradiating the system with pulses from an HCl laser, and so populating the v = 1 level of the reactant HCl [11].

5.3.3
Laser-Sensitised Reactions

Quite a different area of laser-enhanced chemistry involves the sensitisation of reactions by the excitation of a species which does not itself undergo chemical change; this can be regarded as a form of *laser-assisted homogeneous catalysis*, although the term *laser-sensitised reaction* is more often used to describe it. This kind of reaction generally proceeds as a result of the collisional transfer of vibrational energy, often referred to as *V–V transfer*, from molecules of the laser-excited species (the sensitiser) to reactant molecules. The photosensitiser compounds used in connection with photodynamic therapy operate on a very different basis, and are separately discussed in Sect. 5.6.

The major advantage of laser sensitisation becomes apparent if the reactants do not themselves strongly absorb in the emission region of the usual CO_2 laser. By choosing a strongly absorbing sensitiser to initiate the reaction, the rate of reaction induced by laser stimulation can be greatly increased. Both sulphur hexafluoride and silicon tetrafluoride have been widely employed as sensitisers; the latter is nonetheless generally preferred since SF_6 decomposes somewhat too readily under even fairly low-power irradiation. In the presence of SiF_4, various types of sensitised gas-phase reaction have been observed. As illustrated below, these include isomerisation (5.14), condensation (5.15), and retro-Diels-Alder reactions (5.16), (5.17) [12, 13]:

$$CH_2{=}C{=}CH_2 \;\rightarrow\; CH_3C{\equiv}CH \qquad\qquad (5.14)$$

$$2\,CHClF_2 \qquad \rightarrow\; CF_2{=}CF_2 + 2\,HCl \qquad\qquad (5.15)$$

$$(5.16)$$

$$(5.17)$$

Many such reactions which are normally carried out at high temperatures, or even with cw laser heating, produce chemically cleaner products if they are induced indirectly by laser sensitisation since the reaction vessel remains cold. In addition to the factors listed in Sect. 5.2.3, such reactions may also be strongly influenced by the choice of sensitiser and the pressure ratio of sensitiser to reagent.

The combined effects of varying laser fluence and sensitiser pressure are nicely illustrated in some work by Danen and co-workers [14]. This study concerns the free radical chlorination of cyclopropane, a process which when initiated by thermal or conventional photochemical methods generally produces a complex mixture of products which are difficult to separate. A comparison of the product ratios in Table 5.1 shows how marked the effects of changing sensitiser pressure can be; the results also reveal the synthetic advantage under

Table 5.1 Effects of varying sensitiser pressure and carbon dioxide laser fluence (1025.3 cm⁻¹) in the chlorination of cyclopropane, with partial pressures of cyclopropane, chlorine and di-t-butyl peroxide radical initiator 10.0, 2.0 and 0.03 torr, respectively: (a)=chlorocyclopropane, (b)=3-chloroprop-1-ene; (c) 1,1-dichlorocyclopropane, (d)=1,3-dichloropropane + other products [14]

$p(SiF_4)$ (torr)	Fluence (10^4 J m^{-2})	(a) %	(b) %	(c) %	(d) %
1.0	0.62	87.5	0.0	5.1	7.4
1.0	1.22	83.5	2.6	4.0	9.9
2.5	1.24	64.9	7.8	7.8	19.5
5.0	1.22	7.7	64.5	3.8	24.0
Flashlamp-induced[a]		49.5	0.0	49.5	1.0
Heat-induced[b]		29.6	11.6	0.3	58.5

[a] At 298 K;
[b] 0.05 s at 858 K.

suitable conditions of the laser-induced process, compared to non-laser methods, for preparation of the mono-chloro substituted product.

Because the reactants in a sensitised reaction do not need to possess absorption bands in any particular infra-red region, then with a good sensitiser like SiF_4, the range of gas-phase reactions which can be laser-induced is almost limitless. However, there is not a great deal of selectivity in the mechanism for the initial transfer of vibrational energy from the sensitiser to other molecules. This is particularly the case if the reactants are large polyatomics with quasi-continuous vibrational energy levels; here, we cannot in general expect to find isotopic or even isomeric specificity.

Another related topic is *laser-catalysed* reaction, a term which is a very definite misnomer but is applied to a reaction in which the catalyst is itself produced by laser chemistry. For example, the laser pyrolysis of OCS using 248 nm radiation from a KrF laser produces ground state S_2 molecules, which can catalyse the isomerisation of *cis*-2-butene to *trans*-2-butene with an effective quantum yield of about 200 [15].

5.3.4
Ultrafast Reactions

We now turn to a quite distinct branch of laser-induced chemistry concerned with processes induced by ultrashort (picosecond and sub-picosecond) laser pulses. To produce pulses of this duration requires the adoption of mode-locking techniques (Sect. 3.3.3), and, indeed, the era of picosecond measurements dates back to the mid-1960s when the first mode-locked solid-state laser was made operational. The shortest pulses can be obtained with radiation at the uv/visible end of the spectrum, where electronic transitions take place. The excitation processes produced by such pulses are in some senses no different from those which could be induced by longer pulses of the same wavelength and power. However, when combined with suitable monitoring instrumentation (see Fig. 3.18, p. 82), the short pulse duration makes it possible to *observe* what goes on at very short times immediately following the absorption of light-the so-called *primary processes*. This has become a very active area of research, particularly in connection with the photochemistry of large aromatic and pseudo-aromatic molecules in which these primary processes occur over sub-nanosecond times. Many of the compounds which have been studied are, in fact, biochemicals.

Before moving on to consider the specifically chemical processes that can occur, we should first take note of the other physical mechanisms which operate over picosecond and sub-picosecond times. Any cyclic process which occurs at a frequency greater than 10^{12} Hz can of course be complete within a picosecond interval; this includes molecular vibrations and the rotations of some small molecules. Rotations, and *rotational diffusion*, are of particular interest in studies of the liquid state, since they constitute the mechanisms for the most rapid fluctuations in local microscopic structure. The process of rotational *relaxation* following absorption by dyes and other polyatomic molecules usually occurs over tens to hundreds of picoseconds. One means for determining the precise timescale for this type of interaction is to probe the absorption of the sample at short intervals of time after excitation with a picosecond laser pulse. One commonly used technique employs plane polarised light for the excitation, so that only molecules in a suitable orientation can absorb the radiation and proceed to an excited state. Thus, until relaxation processes occur there are disproportionately few ground-state molecules in the correct orientation to absorb from a subsequent probe pulse. This anisotropy can be detected by probing with two laser pulses having orthogonal planes of polarisation, and the timescale for anisotropic behaviour to disappear reflects the timescale for relaxation.

Vibrations come into play in other relaxation phenomena, for example those involving *internal conversion* (e.g. $S_1 \rightarrow S_0$) or *intersystem crossing* (e.g. $S_1 \rightarrow T_1$; see Fig. 2.16, p. 52). Some hundreds of cycles of most molecular vibrations can be completed within a picosecond interval, however; and the timescale of an ultrashort laser pulse is therefore sufficient to allow for an appreciable redistribution of vibrational energy over any polyatomic molecule. Such molecular relaxation processes are thought to play a crucial role in determining the outcome of many photochemical reactions; this is certainly true in many gas phase reactions. *Intermolecular energy transfer* by a direct energy exchange mechanism is another highly significant feature of ultrafast chemistry in the condensed phase. Energy can be fairly efficiently transferred by this mechanism over very short distances in sub-nanosecond time intervals; an exponential dependence on the molecular separation R makes this mechanism of little significance for larger distances. A quite distinct coupling mechanism named after Förster has an R^{-6} dependence and applies over a longer range but is generally associated with longer times.

In condensed phase matter, the bulk vibrations known as optical phonons can also dissipate energy over a similar period; here, another mechanism associated with *dephasing* of the vibrations of different molecules also contributes to picosecond relaxation. The time constant τ for vibrational dephasing is generally somewhat smaller than τ', the time constant for energy dissipation, so it is usually the dephasing that is largely responsible for the relaxation of bulk vibrations. To take a specific example, excitation of the CH stretch of ethanol at 2928 cm^{-1} is associated with an energy dissipation time of approximately 20 ps, compared to a dephasing time of approximately 0.25 ps [16]. This kind of information, obtained from stimulated Raman studies, is obviously only achievable using picosecond laser sources. In passing, it is worth noting that, for liquid nitrogen, the difference between the timescales of the two effects is even more dramatic; at sufficiently low temperatures the vibrational dephasing time is a few picoseconds whilst the dissipation time is measured in seconds!

Finally, fluorescent decay may also in certain rare cases occur on a picosecond timescale, as, for example, in cyanine dyes, phthalocyanines and metalloporphyrins with transition metal centres, and also triphenyl methane dyes. Fluorescence can, however, also be limited to picosecond times if there is a non-radiative process competing with it; it thus provides a convenient and widely used means for monitoring the time development of other ultrafast chemical processes.

Genuine chemistry on the picosecond timescale tends to be limited to relatively simple unimolecular reaction mechanisms. The reason is simply that on the molecular level, the range of movement over such short times is very limited, and bimolecular reactions which depend on the translational motion and collision of large molecules or molecular fragments therefore become largely unimportant. The various instrumental techniques used for monitoring ultrafast processes, which are based on mode-locking and pulse-selection methods described in Chap. 3, mostly involve either resonance Raman scattering (see Sect. 4.5.1) or pump/probe methods discussed in more detail below in Sect. 5.3.5. One good example concerns measurements by Zewail et al. [17] of the rate of photodecomposition of ICN. Using carefully tailored 60 fs pulses of 306 nm laser radiation, these authors have studied the kinetics of the I-CN bond fission which follows the absorption process. The wavelength chosen is such that each laser photon supplies not only enough energy to initiate photodissociation, but also to provide an excess energy which produces a substantial recoil velocity for the iodine and cyanide fragments. By measuring laser-induced fluorescence in the cyanide with a pulsed 388 nm laser probe, it has been shown that the bond cleavage occurs on a timescale of 205 $\pm$ 30 fs. Derivation of this kind of information on precisely how fast a chemical bond breaks has only quite recently become possible, as reproducible laser pulses in the femtosecond region have at last become available.

Some of the most extensive studies in the area of ultrafast chemistry concern the picosecond-timescale unimolecular photodetachment of protons in large organic compounds, as, for example, in the reaction

$$(5.18)$$

which is associated with a time constant of around 10^{11} s^{-1}. *Electron transfer* reactions in solution may also be accomplished over picosecond times; for a 1 M solution, 100 ps would be typical. Certain types of isomerisation not involving the breaking of chemical bonds, such as those involving *cis-trans* conversion, can also take place over this kind of timescale. Here, an example is afforded by the equilibrium between *cis-* and *trans*-stilbene:

$$(5.19)$$

for which the forward reaction in solution or the gas phase also takes place in approximately 100 ps.

As mentioned earlier, some of the most interesting work in the area of ultrafast chemistry concerns elucidation of the primary processes in photobiological phenomena, such as the intricate mechanisms of photosynthesis and vision. Some of the key results established by use of ultrashort pulsed laser experiments are described below. Although only results relating to photosynthesis in *plants* are mentioned here, and space does not permit a thorough discussion, it should be pointed out that the much simpler processes in photosynthetic bacteria have also been very extensively studied using picosecond laser instrumentation, and are now even better understood.

In green plants, there are two distinct reaction sequences known as *photosystems* initiated by the absorption of light. Together, these generate the redox potential for the crucial photosynthetic step in which molecules of water are 'split'. Both photosystems involve chlorophyll molecules, the most abundant form of which is chlorophyll *a*, illustrated in Fig. 5.6. Amongst other distinctions, the two photosystems can be differentiated on the basis of the different absorption wavelengths of the chlorophyll pigments. In both cases there is absorption in the red, with broad bands of half-width around 30 nm in vivo. However, the absorption is centred at around 700 nm in photosystem I, and around 680 nm in photosystem II; the pigments are accordingly referred to as

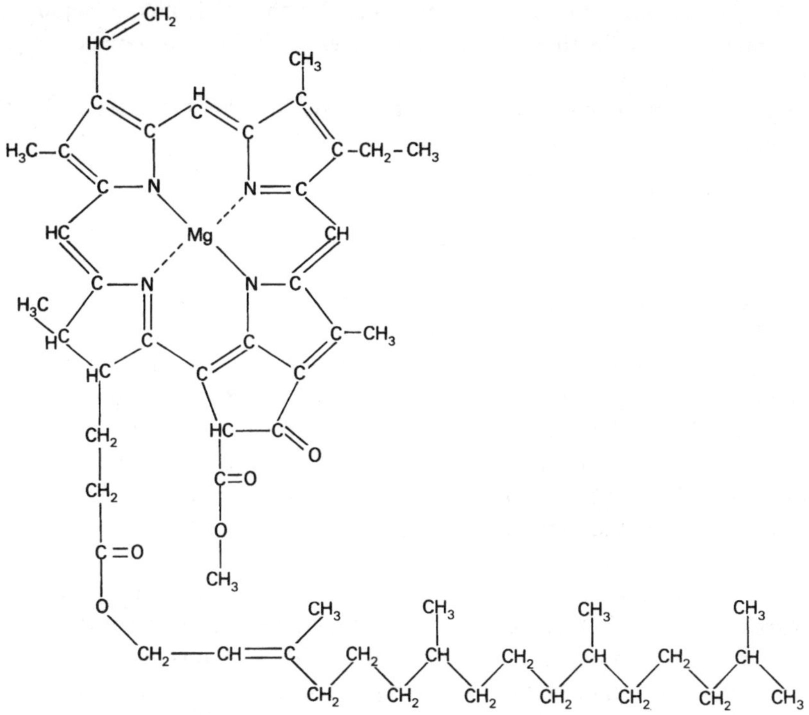

Fig. 5.6 Molecular structure of chlorophyll *a*; the porphyrin ring at the top of the diagram is responsible for the main spectroscopic properties of the molecule

P700 and P680, respectively. In photosystem PS I, the major primary processes can be summarised as follows:

$$\text{Chl } a + h\nu \rightarrow \text{Chl } a^* \tag{5.20}$$

$$\text{Chl } a^* + \text{Chl } a \rightarrow \text{Chl } a + \text{Chl } a^* \tag{5.21}$$

$$\text{Chl } a^* + \text{P700} \rightarrow \text{Chl } a + \text{P700}^* \tag{5.22}$$

$$(\text{P700}^*.\text{A}) \rightarrow (\text{P700}^+.\text{A}^-) \tag{5.23}$$

The photoabsorption step (5.20) represents the initial excitation of the so-called 'antenna pigment' chlorophyll in the chloroplast, and the subsequent equation (5.21) represents rapid Förster transfer of the excitation over a series of chlorophyll molecules. Energy is eventually transferred to the reaction centre (5.22), resulting in electron transfer to an electron acceptor labelled A (5.23). The result of the sequence of reactions (5.20) to (5.23) is formation of a strong reducing agent A^- and a weak oxidant $P700^+$. A similar scheme based on P680 chlorophyll takes place in photosystem II, and results in formation of a strong oxidant $P680^+$ and a weak reductant Y^-. By monitoring the associated changes in spectral characteristics, all of these primary processes have been shown to occur on a picosecond timescale. For the PS II charge separation alone, there prove to be three identifiable stages, with time constants of around 3, 21 and 100 ps.

The subsequent chemistry of photosynthesis may be summarised by the equations:

$$2H_2O \rightarrow 4H^+ + O_2 + 4\,e \tag{5.24}$$

$$CO_2 + 4\,H^+ + 4e \rightarrow (CH_2O) + H_2O \tag{5.25}$$

Since the oxidation of water in Eq. (5.24) and the reduction of carbon dioxide in (5.25) involve four electrons in each of the two photosystems, the production of a single carbohydrate unit (CH_2O) evidently requires the absorption of eight photons. The entire photosynthetic process generally takes place in units containing about 2500 chlorophyll molecules, and the high speed and efficiency of the primary energy and electron transfer processes involving these molecules is clearly of the utmost importance. Much research in this area has been motivated by the desire to create solar energy cells which can mimic this high efficiency in a non-biological system.

Ultrafast reactions also play an important role in the mechanism of vision. Most research work in this area has involved the photochemistry of rhodopsin, a visual pigment composed of over 300 amino acids [18]. It is the absorption of light by rhodopsin molecules in discs contained in the rod cells of the retina

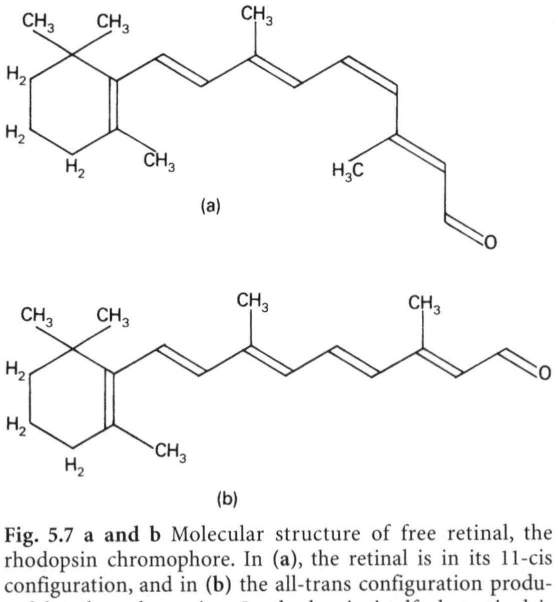

(a)

(b)

Fig. 5.7 a and b Molecular structure of free retinal, the rhodopsin chromophore. In (a), the retinal is in its 11-cis configuration, and in (b) the all-trans configuration produced by photoabsorption. In rhodopsin itself, the retinal is linked to the terminal lysine group of a protein via an aldimine bond

which initiates the basic response of the human eye to light; there are 10^9 such molecules in each cell, and over 10^8 rod cells covering the retina. The absorption of light takes place in a chromophore group known as retinal, an aldehyde derivative of vitamin A displaying broad-band absorption over most of the visible range centred on 498 nm. (It is, incidentally, a molecule which can nicely illustrate application of the *particle in a box* model of quantum mechanics.) Retinal (R498) is present as an 11-*cis* isomer with the remaining linkages thought to be *trans*, as shown in Fig. 5.7a.

Although the detailed mechanism is still not known for certain, it has been established that the primary result of photoabsorption is isomerisation of the retinal to the all-*trans* configuration shown in Fig. 5.7b. The first evidence of chemical change is in fact a shift in the peak of the absorption band to 548 nm, shown using mode-locked laser instrumentation to be complete within 200 fs of the absorption and with a quantum efficiency of about 0.7. One surprise is that the isomerisation essentially occurs on a vibrational timescale, indicating that intramolecular vibrational energy randomisation is comparatively unimportant. In contrast to photosynthesis, the energy of the absorbed photon is here not transferred between pigment molecules; the primary visual processes all take place within the rhodopsin molecule. Subsequently, the activated rhodopsin migrates to the cell membrane, where it results in a change in permeability to sodium ions, and thereby initiates the nerve response.

5.3.5
Laser Reaction Diagnostics

To conclude this section, it is appropriate to recap on some of the major diagnostic methods used for studying the kinetics of laser-induced reactions. Spectroscopic methods based on pulsed lasers are of course particularly appropriate for this purpose, and the relevant principles and instrumentation have mostly been discussed at length in earlier chapters. Many of these techniques were first developed for the study of reaction rates by *flash photolysis* using pulses of about 10^{-4} s duration from conventional flashlamp sources. Laser flash photolysis has largely superseded these earlier methods because it offers the twin advantages of higher intensity and much shorter pulse length. This, of course, not only leads to greater precision but also offers the possibility of investigating much faster reactions such as those discussed in the previous section.

The high sensitivity of many laser-based methods thus makes them ideal for monitoring the concentrations not only of reactant or product species but also for short-lived transient reaction intermediates. For example, many ultrafast reactions can be studied by splitting a single pulse from a mode-locked laser into two parts, one of which is used as a pump to initiate the photochemical reaction, and the other of which is passed through a variable time-delay as shown in Fig. 3.18, p. 82. The delayed part of the pulse can be then be used to generate a probe supercontinuum (see Sect. 3.3.3), and the absorption of different wavelengths by the sample can be used to monitor the appearance and disappearance of various species with different absorption characteristics. Monitoring the variation in spectral response with delay time thus provides very comprehensive kinetic data.

For the majority of laser-induced reactions which occur in the gas phase and over somewhat longer timescales, other methods are more appropriate. For example, the build-up and decay of free radicals is commonly detected by measurement of laser-induced fluorescence using boxcar methods, or by laser magnetic resonance. Examples of the species detected in this way are radicals such as CH, CH_2, C_2 and C_3 which play an important role in many organic reactions, particularly those involving combustion. Species such as C_2, however, have also been shown to appear as transients in the multiphoton dissociation of many organic compounds such as ethane, ethene, and their monosubstituted derivatives. In the case of bimolecular reactions, the activated complex (transition state) species can often be monitored in a similar manner. It is not only the reaction kinetics which can be determined by such methods, however; the richness of information in the spectra can also be used to provide structural information on the nature of the transition state, which thereby facilitates elucidation of the reaction mechanism.

One particular technique for studying laser-induced reactions in the gas phase is worth describing in a little more detail, and that involves the use of supersonic molecular beams. As described earlier (Sect. 4.3.2), a molecular beam apparatus produces a beam of molecules with a very narrow velocity distribution, in which collisional processes are minimised. Such a beam typically has an effective temperature of only a few degrees Kelvin, and usually only the molecular rotational states of lowest energy are populated. The great advantage from the point of view of reaction diagnostics is that by crossing a beam of reactant molecules with a laser beam of the appropriate wavelength, any chosen energy levels can be selectively excited in the reactant. For example, the reactant may be excited to energies above a unimolecular reaction threshold, or even into the ionisation continuum. Any subsequent reaction can then be monitored by interception of the molecular beam by a second, probe laser beam, as shown in Fig. 5.8a. Since the molecular beam travels with a uniform translational speed (typically around 500 m s^{-1}) determined by the source temperature and a velocity selector, the reaction kinetics can be studied by varying the distance between the points of intersection of the molecular beam with the two laser beams.

By inclusion of a second molecular beam of a different species, as shown in Fig. 5.8b, the kinetics of *bimolecular* laser-induced reactions can be studied. This provides the facility for selectively exciting specific vibration or rotation–vibration levels of either reactant prior to reaction. (Usually the molecules involved are comparatively small diatomic or triatomic species, so that intramolecular relaxation processes do not interfere to the extent that they

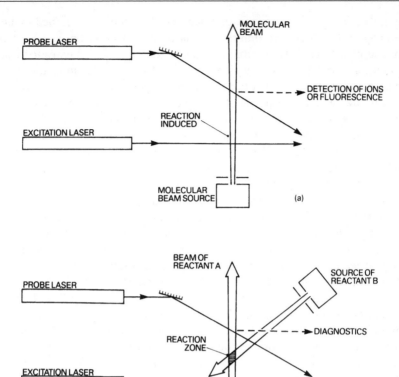

Fig. 5.8a and b Laser reaction diagnostics using molecular beams; (a) for the study of a unimolecular reaction, and (b) a bimolecular reaction

would in larger polyatomics.) The course of reaction is generally monitored using a fluorescence or CARS technique or else by ion detection in a photo-fragment mass spectrometer. The results of such studies provide enormously detailed information and facilitate the determination of rate constants for the elementary reaction steps involved in a complex chemical reaction.

5.4
Isotope Separation

Having examined the principles of laser-induced chemistry, we now move on to discuss one of the major areas of application in isotope separation. The separation of isotopes using lasers is possible by a large number of different methods. Many of these involve photochemical principles examined earlier in the chapter, but there are others such as photodeflection based on different physical principles. It is possible to broadly classify laser schemes for isotope separation into four classes which involve selective ionisation, dissociation, reaction and deflection, each to be discussed below. The common factor in all cases is the selective response of isotopically different compounds (*isotopomers*) to laser radiation, based on the monochromaticity of the radiation, and the isotope-dependence of absorption frequencies. This whole area is one in which there is a great deal of interest and enthusiasm, particularly from the nuclear and associated industries. Whilst with certain compounds this connection is not overt, it is fascinating to observe how many studies have been made of the chemically rather uninteresting molecule SF_6. Of course, this may be related to the fact that uranium hexafluoride has precisely the same structure.

Isotope separation schemes seldom lead to complete separation, but generally effect an enrichment in the relative abundance of a particular isotope. The essential features of any workable isotope separation scheme are a well-resolved isotopic shift in absorption frequency for the starting material, laser radiation which has a linewidth smaller than the extent of this shift, and an efficient extraction stage in which isotopic selectivity is retained. The starting material and the end-product of the process may be chemically quite different, but once an acceptable isotopic abundance has been achieved, further conventional chemical processing can be carried out. The most useful quantitative measure of the efficiency of a separation scheme is the *enrichment factor*, also known as the coefficient of separation selectivity. This is defined in terms of a ratio of the fractional content of the desired isotope in the enriched mixture to the content in the starting material, and is usually given the symbol β.

Suppose, for example, we have a scheme in which the starting material is a compound R, and the end-product is P. If the original material consists of a mixture of two isotopomers R_1 and R_2, and the processing leads to formation of corresponding products P_1 and P_2, then β is given by

$$\beta = \frac{N(P_1)/N(P_2)}{N(R_1)/N(R_2)} = \frac{X(P_1)/(1 - X(P_1))}{X(R_1)/(1 - X(R_1))}, \qquad (5.26)$$

where N represents the number of moles of each species, and X the corresponding mole fraction. It is assumed that the scheme is designed for enrichment of the isotope contained in R_1 and P_1; hence values of $\beta > 1$ are required for successful enrichment. Typical values for laser separation lie between 1 and 10^4. As a concrete example, we can consider the condition for use of a laser enrichment process to produce nuclear fuel from naturally occurring uranium. The mole fraction of U^{235} in natural uranium is approximately 7.1×10^{-3}, and for reactor grade product a mole fraction of 3.0×10^{-2} is required: it is readily calculated using Eq. (5.26) that an enrichment factor of 4.3 is sufficient for this purpose.

5.4.1
Photoionisation

One of the simplest means for achieving laser isotope separation is atomic photoionisation using ultraviolet radiation, as, for example, from an excimer laser. In the conventional process of photoionisation involving single photon absorption, a wavelength usually less than 150 nm has to be employed. The ion and free electron thus formed separate with a kinetic energy determined by the excess of the photon energy over that required to reach the ionisation threshold; see Fig. 5.9a. To avoid recombination of the ion and electron, it is therefore best to use frequencies that provide for transition well up into the ionisation continuum. This, of course, makes the phenomenon completely unselective towards different isotopes. In fact, the wavelength selectivity required for effective isotope separation is almost invariably associated with a transition between states with *discrete energies*.

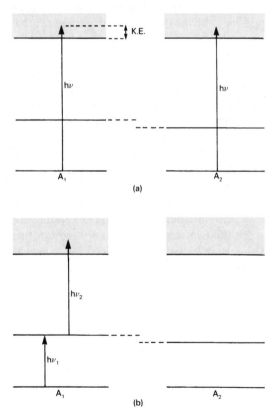

Fig. 5.9a and b Atomic photoionisation processes. In (a), single-photon absorption results in ionisation of both isotopes A_1 and A_2. In (b), two sequential absorptions result in the selective ionisation of A_1

One way over this problem is to produce ionisation by a two-step absorption process, as illustrated in Fig. 5.9b. Here, the primary absorption of a photon with frequency ν_1 produces a transition between two discrete (bound) electronic energy levels, and the isotopic shift is such that only one isotope has the correct energy level spacing to undergo this transition. Subsequent absorption of a second photon of frequency ν_2 produces the required ionisation only in the selected isotope. Both laser frequencies must, of course, be low enough to make direct single-photon ionisation impossible, so that we require

$$h\nu_1, \; h\nu_2 < I < h\nu_1 + h\nu_2, \tag{5.27}$$

where I is the ionisation energy. Since the frequency of the photon absorbed in the first step is precisely determined by the energy level spacing in the appropriate isotope, a tunable laser is often employed to produce the ν_1 radiation. There is, however, relative freedom over the choice of frequency of the secondary, ionising transition. It is therefore often possible to adopt the simplest solution and arrange for both of the absorbed photons to have the same frequency. In this way, only a single laser beam need be involved.

This method of isotope separation is generally practised by the laser irradiation of a beam of neutral atoms, with removal of the ions so formed by a strong electric field of the order of $10^5 \, \text{V m}^{-1}$. Using this type of scheme, enrichment factors in the range 10^2–10^3 have been successfully achieved. Following a very large research effort into this method of isotope separation, it has been selected by the US Department of Energy for the next generation of uranium enrichment plants. The specific principle involved in this application is the selective ionisation of ^{235}U atoms using a dye laser pumped by a powerful 150-W copper vapour laser, in apparatus of the kind illustrated in Fig. 5.10. It is possible to obtain a similar separation effect in isotopomeric *molecules*, although in this case photodissociation is more commonly used. There are also several variations on the two-step sequential absorption theme; for example, the second transition may be collision-induced, or it may populate a bound state that subsequently autoionises, or else more than two steps may be involved.

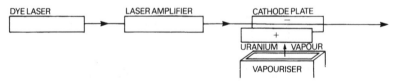

Fig. 5.10 Schematic apparatus for atomic vapour laser isotope separation based on the principle of photoionisation. The uranium vapour contains both ^{235}U and ^{238}U, but only the ^{235}U atoms are ionised by the tuned dye laser beam, and the ions thus formed are collected electrically on a flat cathode plate. The ^{238}U passes on to a separate collector

5.4.2
Photodissociation

There are a number of ways in which isotopes can be separated by the selective photodissociation of isotopomers. Generally, these take advantage of the comparatively large isotopic shifts associated with vibrational energy levels. As with photoionisation, direct single-photon absorption of uv/visible light is by no means ideal for the purpose of isotope separation, since it does not couple two states with discrete energy. Exceptions to this arise in molecules exhibiting *predissociation*, as illustrated in Fig. 5.11. Here, dissociation from a repulsive electronic excited state can be accomplished following a single-photon transition which involves only *discrete* vibrational levels, i.e. those belonging to the electronic ground state and a *bound* electronic excited state. A good example is provided by formaldehyde, whose first excited singlet state exhibits predissociation at wavelengths below 354.8 nm. In a mixture of formaldehyde isotopomers containing hydrogen and deuterium, absorption of 325.0 nm radiation from a He-Cd laser preferentially produces predissociation of the HDCO:

$$HDCO \rightarrow \begin{cases} HD + CO \\ H \quad + DCO \\ D \quad + HCO \end{cases} \qquad (5.28)$$

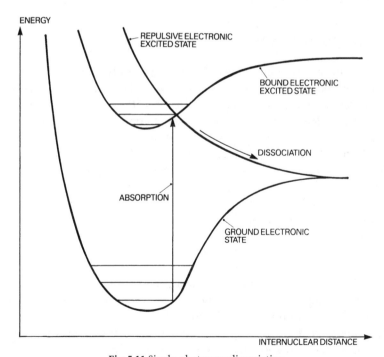

Fig. 5.11 Single-photon predissociation

The HD product can then be thermally reacted with oxygen to form HDO, and subsequent fractional distillation yields D_2O.

Generally speaking, however, isotopically selective photodissociation necessitates absorption of more than one photon and provides us with several alternatives. Two-step absorption is an obvious option, where the first step involves an isotope-selective transition between discrete states, and dissociation is induced by absorption of a second photon. The primary absorption may produce vibrational excitation within the ground electronic state or it may populate a vibrational level of a bound electronic excited state. The former arrangement as illustrated in Fig. 5.12 is more common and generally requires irradiation with both infra-red and ultra-violet laser light. One condition for satisfactory operation of this method is to minimise *thermal* population of the vibrational levels. The first experiments on ammonia using this method well illustrated its potential [19]. In a mixture of $^{14}NH_3$ and $^{15}NH_3$, irradiation with the P(16) 10.6 μm line (947.7 cm^{-1}) from a TEA carbon dioxide laser produces selective vibrational excitation of $^{15}NH_3$, and subsequent dissociation can be induced by any suitably intense ultraviolet source. The sequence of reactions is as follows:

$$^{15}NH_3 + h\nu_1 \rightarrow {}^{15}NH_3^{\ddagger} \tag{5.29}$$

$$^{15}NH_3^{\ddagger} + h\nu_2 \rightarrow {}^{15}NH_2 + H \tag{5.30}$$

$$^{15}NH_2 + {}^{15}NH_2 \rightarrow {}^{15}N_2H_4 \tag{5.31}$$

$$^{15}N_2H_4 + H \rightarrow {}^{15}N_2H_3 + H_2 \tag{5.32}$$

$$^{15}N_2H_3 + {}^{15}N_2H_3 \rightarrow 2{}^{15}NH_3 + {}^{15}N_2 \tag{5.33}$$

Since none of the elementary steps following absorption involve molecules of ammonia itself, the $^{14}NH_3$ does not enter into the reaction scheme, and so isotopic selectivity is retained; an enrichment factor of about 4 is typical here.

The most widely studied method of laser isotope separation is undoubtedly multiphoton infra-red dissociation, based on the principles discussed in Sect. 5.2 and illustrated by Fig. 5.2. Here, 30 or 40 photons may be involved in the process of excitation, but isotopic selectivity applies only over energy region I where the first few photons are absorbed. Once more, this selectivity results from the relatively large differences between corresponding vibrational energy levels of isotopomers. The enrichment factor usually reaches its highest values at low temperatures and fairly low pressures in the millibar range where collision broadening is minimised. In large molecules, even if there are discrete vibrational frequencies with suitably large isotopic shifts, the redistribution of vibrational energy amongst the various modes following absorption reduces the isotopic selectivity. Hence, most successful results based on this principle have involved fairly small molecules.

Here, there are a great many case studies, the majority making use of the intense infra-red radiation from a carbon dioxide laser. Using the P(30) (934.9 cm^{-1}) line from the 10.6 μm band, for example, it has been shown that

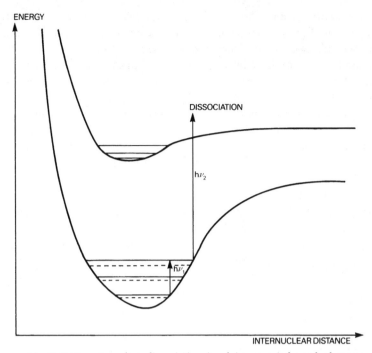

Fig. 5.12 Two-step photodissociation involving one infra-red photon (hv_1) and one ultraviolet photon (hv_2). The vibrational levels of another isotopomer of the same compound are indicated by the dotted lines

dissociation of a CF_3CTBrF/CF_3CHBrF mixture provides very high selectivity for the tritium isotopomer [20]. The most widely studied case is the separation of the sulphur isotopes ^{32}S and ^{34}S, based on selective multiphoton dissociation of SF_6. For example, irradiation at around 10.59 μm selectively excites the ν_3 fundamental vibration of $^{32}SF_6$, and leads to its selective dissociation. The magnitude of the enrichment factor in such studies depends on several factors, including pressure and laser intensity, and can be either positively or negatively influenced by the presence of other gases. Unfortunately, the prime candidate for laser isotope separation, uranium hexafluoride, does not absorb in this wavelength region, and much effort has been directed into finding a suitable laser source operating in the appropriate 16 μm region. Tunable diode lasers which operate at this wavelength have insufficient power for this purpose, but can nonetheless be used to measure the concentration of fissionable ^{235}U.

5.4.3
Photochemical Reaction

We now turn to a consideration of potentially isotope-selective photochemical reactions in which the initial absorption of light does not, in contrast to the processes examined above, directly lead to ionisation or fragmentation. In general, the result of photoabsorption in such cases is the production of an excited state which subsequently undergoes a chemical reaction.

The simplest unimolecular reaction which can take place is isomerisation. Provided this leads to formation of a *geometrical* isomer, in which there are changes in the chemical bonding, the chemical and physical properties of the product and the starting material will generally be quite different, and their separation is a simple matter. The requirement is then for an isotopically selective process of *photoisomerisation*. Once again, excitation of vibrational levels provides the best means of obtaining isotopic selectivity, and the carbon dioxide laser is the favourite infra-red source. An example is afforded by methyl isocyanide, CH_3NC, which isomerises to methyl cyanide, CH_3CN, by strongly pumping at various CO_2 laser wavelengths; the associated enrichment factors for ^{13}C and ^{15}N are in the neighbourhood of 1.2.

Laser-induced reactions involving more than one reactant usually consist of the photoexcitation of one reactant, followed by its reaction with a second species:

$$X + h\nu \rightarrow X^* \tag{5.34}$$

$$X^* + Y \rightarrow Z \tag{5.35}$$

where the species X^* may in fact be either electronically or vibrationally excited. If the absorption process (5.34) is the isotope-selective step, an obvious requirement for retention of specificity is that Y reacts preferentially with X^* rather than X. A useful example is the reaction between iodine monochloride and bromobenzene. Irradiation of the mixture at 605.3 nm selectively excites $I^{37}Cl$, which subsequently undergoes the reaction

$$I^{37}Cl^* + C_6H_5Br \rightarrow I + C_6H_5^{37}Cl + Br \tag{5.36}$$

The chlorobenzene product of the reaction has a ^{37}Cl content well above the level of natural abundance. The isotopic selectivity of the reaction is mainly limited by competing processes such as the collisional transfer of energy from $I^{37}Cl$ to $I^{35}Cl$. It is important to note that the main reaction is *not* one of ICl dissociation followed by radical reaction with C_6H_5Br; the photon energy initially absorbed by the ICl is only sufficient to provide transition to a *bound* excited state $(A^3\pi_1)$.

5.4.4
Photodeflection

The last of the principal laser processes used to obtain isotope separation involves photodeflection and is based on the concept of *radiation pressure*. Every photon carries a momentum of magnitude h/λ which is imparted to any atom or molecule by which it is absorbed. A photon of visible light, for example, has an associated momentum of the order of 10^{-27} kg m s^{-1}, and its absorption by a free atom thus typically increases the atomic velocity by approximately 1 cm s^{-1}. In an atomic beam containing more than one isotope, laser irradiation at a wavelength corresponding to absorption in one particular isotope results in its selective excitation. Thus, if the laser beam and the atomic beam intersect perpendicularly, this produces a deflection of the appropriate atoms away from the beam direction.

Since a typical atomic beam velocity is 500 m s^{-1}, the angular deflection of $(1\,\text{cm s}^{-1}/500\,\text{m s}^{-1}) = 2 \times 10^{-5}$ rads is minimal. However, since the lifetime of electronic excited states is typically measured in nanoseconds, the rapid decay back to the ground state enables a number of photons to be successively absorbed by each atom during the time it takes to traverse the laser beam. Each intervening radiative decay also involves momentum transfer, but this does not cause any net effect since emission is in random directions. Nonetheless, to produce an angular deflection of 10^{-3} rads, which even then corresponds to a displacement of only 1 mm over a 1-m distance, requires each atom to be excited something like fifty times as it traverses the laser beam. In energy terms, this is rather inefficient, and corresponds to an input of about 10 MJ mol^{-1}. Successful experiments based on this principle have, nonetheless, been carried out with barium, which in its natural occurrence contains isotopes of mass number ranging from 134 to 138. In this case a dye laser tuned to 553.5 nm, for example, can be used to selectively excite and separate the ^{138}Ba isotope.

Before leaving this topic, it is worth drawing attention to another aspect of the same principle of resonant absorption. This concerns the subject of *laser cooling and trapping*. It has been shown [21, 22] that by irradiating a beam of sodium atoms with a counterpropagating beam from a ring dye laser tuned to one of the hyperfine components of the 3s $^2S_{1/2} - $ 3p $^2P_{3/2}$ transition, (the sodium D-line), each atom can effectively be stopped in its tracks by the successive absorption and emission of about 20 000 photons, reducing the kinetic temperature of the beam to well below 0.1 K. In this case, the laser frequency has to be continually varied to compensate for the gradual removal of the Doppler shift in absorption frequency as the atomic beam is progressively slowed down. This technique represents a new technology which should open some very interesting research possibilities for the study of isolated atoms.

5.5
Materials Applications

5.5.1
Laser Surface Chemistry

Surface chemistry is an increasingly significant discipline in which lasers are employed. A large number of the chemical applications of lasers to surfaces concern either spectroscopic methods dealt with earlier (see, for example,

Sect. 4.5.5) or other surface-enhanced optical processes. From the point of view of laser-induced chemistry, many of the most important topics in this field concern the treatment of semiconductor surfaces and therein hold enormous potential for application in the manufacture of microelectronic devices. It is worth noting that excimer lasers in particular produce emission in a very useful wavelength range, where photon energies are sufficient to break chemical bonds in a variety of compounds involving the Group IV elements. Because a high level of attenuation is associated with the corresponding absorption, surface treatment is a particularly obvious application for these lasers, and the high power levels they produce are such that the rate of processing can be viable for production purposes.

Here we shall concentrate specifically on laser-induced chemical *reactions*, though in this area it is often quite hard to draw a clear line between chemical and physical processes. Techniques such as laser-induced surface alloying and cladding may be regarded as occupying the middle ground, whilst laser machining, annealing, recrystallisation and other essentially physical processes also contribute to the wide-ranging potential for laser materials processing. For a comprehensive treatment of these topics the reader is referred to the excellent monographs by Bäuerle and Steen (details in Appendix 3). In many such applications, as well as those to be considered below, the simplest alternative to the use of a laser is heat treatment. One of the principal advantages of using a laser, however, is that its tightly focussed beam allows the treatment of very small areas of surface down to 1 μm or less in diameter without affecting the surrounding material. This in part accounts for recent interest in the enhanced electroplating that occurs when laser light is focussed on an absorbing cathode, rates of metal deposition from the electrolyte being increased by a factor of up to a thousand in the irradiated area.

As a first example of laser-induced surface *chemistry*, we can consider multiphoton dissociation reactions in the gas phase, where the surface plays the role of a heterogeneous catalyst. Here, the course of reaction is influenced by other factors in addition to the four listed in Sect. 5.2.3. For example, in the dissociation of propan-2-ol over CuO using 1070.5 cm^{-1} radiation from a CO_2 laser [23], there are two competing reaction pathways leading to different products:

$$CH_3CH(OH)CH_3 \begin{cases} CH_3\overset{\overset{\text{O}}{\|}}{C}CH_3 + H_2 \\ \\ CH_2{=}CHCH_3 + H_2O \end{cases} \qquad (5.37)$$

Here it is found that the product ratio of propanone/propene can be varied from 0.02 to 6, depending on the orientation of the catalytic surface relative to the laser beam.

Much laser-induced surface chemistry involves the principle of depositing a thin film covering onto a substrate surface by decomposition of a gas. This method is known in the jargon of the laser field as *laser chemical vapour deposition* and represents an alternative to the high-temperature methods more usually employed for the construction of microelectronic devices. The mask-free writing of an adsorbate onto semi-conductor surfaces by laser deposition provides a classic illustration of an application facilitated by the distinctive properties of laser light, but which would be greatly more difficult by traditional thermal methods. It has, in fact, already been proved possible to fabricate viable integrated circuits using the laser method. In the case of laser deposition, although the chemistry of interest actually concems the vapour, reaction takes place principally at or near to the point at which the substrate is irradiated. The principle involved in the process of deposition may be either pyrolytic or photolytic by nature. For both types of deposition, laser irradiances are typically of the order 10^{12} W m^{-2}, and the partial vapour pressure of the vapour in the range $10^{-3} - 1$ atm. Under these conditions, rates of deposition with a scanning laser beam are typically between 0.1 and 100 μm s^{-1}.

Pyrolytic deposition involves thermal reaction and is, in general, an indirect result of the surface heating produced by the laser radiation. For this purpose, it is obviously necessary for the substrate to absorb in the appropriate wavelength region. For example, amorphous films of silicon can be pyrolytically deposited from SiH$_4$ vapour onto quartz or various other surfaces irradiated by 10.59 μm radiation from a carbon dioxide laser, using apparatus such as that shown in Fig. 5.13. In this particular case, the silane can itself absorb the radiation, and it is considered likely that multiphoton dissociation also plays a part in the dehydrogenation process by causing molecular fragmentation before surface deposition occurs. However, even when the chemical mechanism involved is purely a thermal one, the laser method still has several advantages over the traditional heat-induced deposition process, since it leads to higher rates of deposition and much more precisely controlled localisation of the surface coverage.

Photolytic deposition (*photodeposition*), by contrast, results directly from the absorption of laser light by molecules of the vapour. This is a technique offering a great deal more control and selectivity, since different compounds will absorb at different wavelengths. Once more, the localisation of reaction is

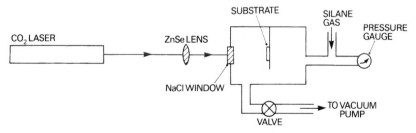

Fig. 5.13 Apparatus for the laser chemical vapour deposition of silicon by the pyrolytic dissociation of silane

an additionally attractive and significant feature of the laser-induced process. Very often an excimer laser or else the frequency-doubled light from an ion laser provides the ultraviolet photon energies necessary to produce dissociation. There have been numerous studies concerning deposition on semiconductor surfaces, and a good example is afforded by the photolysis of dimethyl cadmium, $Cd(CH_3)_2$. This compound decomposes to form a surface coating of pure cadmium metal and represents one of an increasingly large number of organometallic compounds that can be used for the deposition of metals. It is also important to notice that laser methods can be used for *doping* semiconductors, for example by the photolysis of BCl_3 or PCl_3; here the laser beam has the additional role of melting the substrate so that the dopant boron or phosphorus atoms can be incorporated into the surface by liquid diffusion.

By contrast, quite a different kind of laser-induced reaction takes place on semiconductor surfaces under bromomethane, since the laser photolysis of CH_3Br releases Br atoms which can subsequently etch the surface. Etching of this kind can be achieved with various organohalogen compounds in the gas phase and can also be accomplished in solution by electrochemical means. Laser-induced chemical etching is in itself an area which shows enormous potential for the microelectronics industry, since in contrast to the more usual plasma etching methods, it does away with the need for any kind of resist mask, and the rate of etching can be faster by a factor of 50 or more.

Another neat example of laser surface chemistry is afforded by studies demonstrating the possibility of laying down an InP layer by co-deposition of indium and phosphorus from a mixture of $(CH_3)_3InP(CH_3)_3$ and $P(CH_3)_3$ [24]. In this case, using 193 nm radiation from an ArF excimer laser, the photodecomposition reactions are:

$$(CH_3)_3InP(CH_3)_3 + 2h\nu \rightarrow In^* + 3CH_3 + P(CH_3)_3 \tag{5.38}$$

$$P(CH_3)_3 + 3h\nu \rightarrow P^* + 3CH_3 \tag{5.39}$$

Other examples of laser-induced surface reaction include the formation of oxide layers by the photo-oxidation of metals and semiconductors. A widely studied example is the formation of SiO_2 on the surface of silicon by photooxidation in an oxygen atmosphere. The enhancement of the rate of oxidation resulting from laser irradiation of the surface is in this case thought to be a result of the breakage of Si–Si bonds through the absorption of the laser radiation. Metal silicides can also be formed by a direct *thermally* induced reaction between a surface film of metal and a silicon substrate.

5.5.2
Purification of Materials

Closely related to the isotope separation methods examined in Sect. 5.4 is a more general area of applications for lasers, that of material purification. Once again, the underlying principle is the specific excitation of a single chemical component in a mixture, in this case usually the impurity. In principle any of the methods discussed in Sect. 5.4 can be applied for this purpose. However, for practical application on an industrial scale, methods based on separation of atoms are clearly inappropriate and photochemical reaction is the only realistic option. An illustration is afforded by the removal of contaminants from silane, SiH_4. This gas is used in the manufacture of silicon-based semiconductor devices, and its purity is crucial for device performance. Using an ArF laser operating at 193 nm, it has been shown that impurities of arsine AsH_3, phosphine PH_3, and diborane B_2H_6 can all be photolysed and so removed from silane gas very effectively; for example, 99% of the arsine can be destroyed at the loss of only 1% of the silane [25]. Another example based on the argon fluoride laser is the removal of H_2S from synthesis gas (a CO/H_2 mixture obtained by coal gasification). This is particularly significant since H_2S readily poisons the catalysts used for hydrocarbon synthesis. One final case which is, again, of relevance to the microelectronics industry is the removal from BCl_3 of carbonyl chloride, $COCl_2$, which is often a fairly troublesome contaminant, using the CO_2 laser. Applications of this type may prove to be economically viable where there is no cheap alternative based on conventional chemical or physical methods.

5.5.3
Production of Ceramic Powders

Another area in which laser methods have proven potential is in the development of new methods for the chemical synthesis of ceramic powders. These are inorganic substances with a high degree of thermal stability, which show much promise for applications in mechanical, electronic and chemical engineering. A typical example is silicon nitride, Si_3N_4, which amongst other commercial methods can be formed by the gas-phase reaction of silane and ammonia:

$$3\ SiH_4 + 4\ NH_3 \rightarrow Si_3N_4 + 12H_2 \tag{5.40}$$

This reaction requires a large input of energy and is therefore usually carried out in a furnace or else in the presence of an electric arc. For many practical applications, however, the distribution of particle sizes in the end-product is a crucial factor, and one for which the conventional methods of production fail to be completely satisfactory.

This problem can now be overcome by the use of radiation from a carbon dioxide laser to produce a vibrationally enhanced bimolecular reaction (see Sect. 5.3.2). Both silane and ammonia have several absorption features in the 10.6 μm wavelength region, and both can therefore undergo vibrational excitation. With beam intensities approaching 10^7 W m^{-2}, the reaction yield is close to 100%. In fact, this particular reaction is especially interesting since it represents one of the first cases of laser synthesis for which the yield is higher and production cost lower than in the traditional chemical method. In one of the possible configurations, a mixture of the reactant gases traverses a laser beam of a few millimetres diameter, and the particles of silicon nitride powder so produced are carried by the gas to be collected at a filter.

The very short and well-controlled exposure time associated with this method generally results in the formation of a very fine powder with an exceptionally narrow distribution of particle sizes, typically less than a micron across. These characteristics maximize the mechanical strength of articles made by compacting the powder, and they also result in a high surface/volume ratio, which is highly significant for catalytic applications. In particular, laser pyrolysis provides the means for production of a range of non-oxide-based catalysts not hitherto available with such large surface area. Colloidal suspensions of such fine powders can also be prepared; these can be made indefinitely stable, and as such may prove useful for doping semiconductors.

5.5.4
Laser-Initiated Polymerisation

One last application of lasers in the chemistry of materials concerns their use for initiating polymerisation reactions. Here it is primarily pulsed uv radiation that is employed to produce radicals for the process initiation. It generally proves that there are substantial differences in the character of polymers obtained with laser radiation, compared to those produced with radiation of the same wavelength and total energy from other sources. One reason is that the high intensities associated with laser radiation can, by increasing the transient concentrations of radical intermediates, substantially increase the extent to which sequential absorption processes enter into the reaction. A second reason is more directly connected with the pulsed nature of the radiation. The radical chain propagation responsible for linking successive monomer units proceeds between pulses, subject to the normal radical decay kinetics, only until the onset of a succeeding pulse. Then, the sudden increase in initiator radical concentration leads to radical–radical termination processes that prevent further chain lengthening. At simplest, the mean chain length in the laser-produced polymer is then directly proportional to the 'dark time' between pulses. So, the product is characterised by a molecular weight distribution more directly amenable to control and generally quite different from the polymer produced using conventional photoinitiation. An increase in the mean molecular weight and narrowing of the distribution often signifies a product which is tougher and which displays better tensile and thermal properties.

The effect of laser pulse frequency on polymer composition is well illustrated by the series of *gel permeation chromatograms* shown in Fig. 5.14 [26]. These curves essentially portray the molecular weight or chain length distribution of various samples of poly(methylmethacrylate), formed by laser initiation with different pulse intervals. In each trace the lower-molecular-

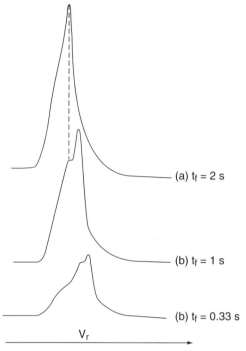

Fig. 5.14 Gel permeation chromatograms for poly(methylmethacrylate) produced by pulsed laser initiation
with various pulse intervals

weight components, which come through the gel first, show as the leading edge
to the right, the bulkier components giving the trailing edge. With a two-second pulse interval (top trace) relatively few radicals persist for the full extent of
the dark time and a nice sharp distribution is obtained. A 1-s interval (middle
trace) gives a higher yield of short, lower-molecular-weight chains, though
radicals that live beyond the succeeding pulse generate some polymer units
with the same chain length as characterises the two-second reaction. When
the pulse interval is one-third of a second (bottom trace), many radicals live
to see two or more pulses and the polymer distribution increasingly broadens
towards the usual (cw irradiation) form. It may be concluded that a product
with better mechanical and thermal properties results when a pulse interval of
2 s or more is employed. Other studies on this and related processes have
shown that products formed by initiating polymerisation in monomer adsor-
bates are also significantly affected by the physical properties of the substrate.

5.6
Photodynamic Therapy

We conclude by looking at one last chemical application of lasers which, at least for certain individuals, may prove to have by far the most profound significance. This concerns a treatment for cancer, based on the intrinsic biological and photochemical properties of certain *photosensitiser* compounds. The best known of these is a haematoporphyrin derivative, usually abbreviated to HpD; many others are also porphyrin-related substances, such as chlorophyll itself (see Fig. 5.6, p. 189) or substituted phthalocyanines of the kind shown in Fig. 5.15 [27]. Mostly these are compounds that absorb at the red end of the spectrum where radiation can penetrate furthest in human tissue. When such a photosensitiser is introduced into the body of a cancer patient, it can be selectively retained in malignant tumours in concentrations substantially greater than those of normal tissue, normally over a period of 2–3 days. Its characteristic fluorescence can serve as a useful means of locating small internal tumours. Moreover, when red light (with a wavelength around 630 nm in the case of HpD) from a suitably powerful laser is directed onto the cancerous tissue, it is strongly absorbed by the photosensitiser dye. In general, for a photosensitiser P, we then see the following sequence:

$$P(S_0) + h\nu \rightarrow P^*(S_1) \tag{5.41}$$

$$P^*(S_1) \rightarrow P^*(T_1) \tag{5.42}$$

$$P^*(T_1) + O_2(T_0) \rightarrow P(S_0) + O_2^*(S_1) \tag{5.43}$$

$$O_2^*(S_1) + \text{tissue} \rightarrow \text{necrosis} \tag{5.44}$$

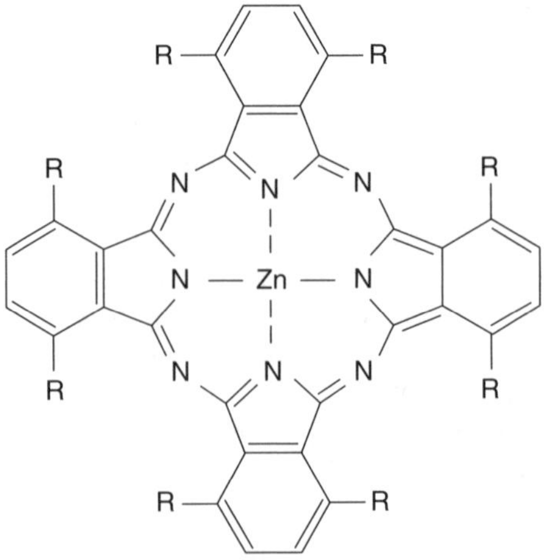

Fig. 5.15 Photosensitiser derivatives of octoalkyl zinc phthalocyanine. R is a straight-chain alkyl group

Photoexcitation (5.41) first populates an electronically excited singlet state of P, which converts to a triplet state (5.42) by intersystem crossing. Energy transfer to the ground state of molecular oxygen, also a triplet, generates highly reactive singlet oxygen (5.43), which then oxidises and destroys cells (5.44) where the photosensitiser is present. Since the diffusion distance of singlet oxygen in vivo is typically only a small fraction of a micron, the tumour is targeted with minimal damage to surrounding healthy tissue. In early clinical trials the gold laser, emitting several watts at 628 nm, proved an ideal source for this type of treatment; as new and more fully characterised photosensitisers have become available, the use of dye lasers with an argon ion pump has become more common, and there is currently a drive towards a viable diode laser system. Though cost considerations have been moving clinical research towards the deployment of non-laser sources, lasers retain a distinct advantage for internal procedures in facilitating optical fibre delivery of the excitation.

5.7
References

1. Rak J, Błażejowski J, Lampe FW (1995) J. Photochem. Photobiol. A90:11
2. Orel AE, Miller WH (1978) Chem. Phys. Letts 57:362
3. Zitter RN, Koster DF (1978) J. Am. Chem. Soc. 100:2265
4. Ambartzumian RV, Chekalin NV, Doljikov VS, Letokhov VS, Lokhman VN (1976) Opt. Commun. 18:220
5. Yoger A, Loewenstein-Benmair RMJ (1973) J. Am. Chem. Soc. 95:8487
6. Yoger A, Benmair RMJ (1977) Chem. Phys. Letts 46:290
7. Malatesta V, Willis C, Hackett PA (1981) J. Am. Chem. Soc. 103:6781
8. Danen WC, Jang JC (1981) In: Steinfeld JI (ed) Laser-induced chemical processes, Plenum, New York
9. Kurylo MJ, Braun W, Xuan CN (1975) J. Chem. Phys. 62:2065
10. Krasnopyorov LN, Chesnokov EN, Panfilov VN (1979) Chem. Phys. 42:345
11. Arnoldi D, Wolfrum J (1974) Chem. Phys. Letts 24:234
12. Cheng C, Keehn P (1977) J. Am. Chem. Soc. 99:5808
13. Garcia D, Keehn P (1978) J. Am. Chem. Soc. 100:6111
14. Danen WC, Setser DW, Nguyen HN, Ibrahim MSB (1987) Spectrochim. Acta. 43A:173
15. Clark JH, Leung KM, Loree TR, Harding LB (1978) In: Zewail AH (ed) Advances in laser chemistry, Springer, Berlin Heidelberg New York
16. Laubereau A, von der Linde D, Kaiser W (1972) Phys. Rev. Letts 28:1162
17. Rosker MK, Dantus M, Zewail AH (1988) Science 241:1200
18. Hargrave PA (1982) Prog. Retinal Res. 1:1
19. Ambartzumian RV, Letokhov VS, Makarov GN, Puvetskii AA (1973) Sov. Phys. JETP Letts 15:501
20. Takeuchi K, Kurihara O, Mahide Y, Midorikawa K, Tashiro H (1985) Appl. Phys. B 37:67

21. Prodan J, Migdall A, Phillips WD, So I, Metcalf H, Dalibord J (1985) Phys. Rev. Letts 54:992
22. Ertmer W, Blatt R, Hall JL, Zhu M (1985) Phys. Rev. Letts 54:996
23. Farneth WD, Zimmerman PG, Hogenkamp DJ, Kennedy SD (1983) J. Am. Chem. Soc. 105:1126
24. Donnelly VM, Geva M, Long J, Karlicek RF (1984) Mat. Res. Soc. Symp. Proc. 29:73
25. Clark JH, Anderson RC (1978) Appl. Phys. Letts 32:46
26. Davis TP (1994) J. Photochem. Photobiol. A 77:1
27. Cook MJ, Chambrier I, Cracknell SJ, Mayes DA, Russell DA (1995) Photochem.Photobiol. 62:542

5.8
Questions

1. A conventional infra-red source of radiation passed through a monochromator and a mechanical chopper typically produces pulses at a wavenumber of 10^4 cm^{-1} with a FWHM bandwidth of 10^{-2} cm^{-1}, a pulse length of 1 ms, and a pulse energy of 10^{12} photons. Give rough estimates of the corresponding values for the carbon dioxide laser and explain why it is therefore a far better pump for selective photochemistry.

2. The carbon dioxide laser is far more widely employed in laser chemistry than any rare gas ion laser. Explain why this is so, by comparison of (a) the relative output characteristics, and (b) the types of process which each laser can photoinitiate. Explain also why completely bond-selective chemistry is seldom achieved with laser excitation.

3. Explain why the photodissociation of CF_3Cl using 10.6 μm radiation from a carbon dioxide laser produces a yield that increases with temperature.

4. "Laser-induced chemistry often involves fewer undesirable side-reactions than are found on simple application of heat, because the reaction takes place in the cold". Comment on the reasoning behind the above statement and explain why the concept of temperature is, in any case, rather ill-defined for a laser-induced gas-phase reaction.

5. The infra-red multiphoton dissociation of CF_3I can be effected using a 9.6 μm line from a Q-switched carbon dioxide laser. This wavelength matches one of the C-F stretching vibrations in CF_3I; the C-I stretching mode is associated with absorption at very much longer wavelengths. However, the principal laser-induced reaction is represented by the reaction $CF_3I \rightarrow CF_3 + I$. Answer the following questions.
   (a) Explain how and why the C-I bond is preferentially broken.
   (b) Calculate the theoretical minimum number of laser photons required to dissociate a single CF_3I molecule.
   (c) Using 15-ns laser pulses, the dissociation is observed to require about 50 photons. Why does this exceed the theoretical minimum?

(C-I bond energy $= 213\,kJ\,mol^{-1}$; $c = 3.00 \times 10^8\,m\,s^{-1}$; $h = 6.63 \times 10^{-34}\,J\,s$; $L = 6.02 \times 10^{23}-1$).

6. The primary process involved in the photochemistry of vision in the retina is a *cis-trans* isomerisation of a chromophore group known as retinal. This is associated with a shift in the visible absorption maximum from 498 to 548 nm. It is known that the retinal isomerisation takes place within a pico-second of photoabsorption; suggest suitable laser instrumentation for analysis of the detailed kinetics.

7. The decay kinetics of a transient molecular species formed as the result of laser flash photolysis can be studied by monitoring its absorption from a probe pulse of light, subsequent to its excitation. The method involves selecting single pulses from a mode-locked train, together with the use of a stepping motor to delay the probe pulse. Draw a schematic diagram illustrating the instrumentation.

8. A laser isotope separation scheme produces an enrichment factor $\beta = 5.0$ for ^{235}U. Given that for reactor grade fuel a mole fraction of 3.0×10^{-2} is required, calculate the minimum mole fraction of ^{235}U which must be present in the untreated uranium.

Listing of Output Wavelengths from Commercial Lasers

The table overleaf lists in order of increasing wavelength λ the emission lines of the most commonly available discrete-wavelength lasers over the range 100 nm–10 µm. Although continuously tunable lasers are not included, the molecular lasers which can be tuned to a large number of closely spaced but discrete wavelengths are listed at the end of the table. Harmonics are indicated by $\times 2$, $\times 3$ etc. Other parameters such as intensity and linewidth vary enormously from model to model, and no meaningful representative figure can be given. However, the annually updated 'Laser Focus Buyers' Guide' and 'Photonics Buyers' Guide' both have comprehensive data on all commercially available lasers, together with manufacturers' addresses.

λ/nm	Laser	λ/nm	Laser	λ/nm	Laser
157.5	Fluorine	437.1	Argon	628.0	Gold
157.6	Fluorine	441.6	Helium-cadmium	632.8	Helium-neon
173.6	Ruby ×4	454.5	Argon	647.1	Krypton
193	Argon fluoride	457.7	Krypton	657.0	Krypton
213	Nd:YAG ×5	457.9	Argon	676.5	Krypton
222	Krypton chloride	461.9	Krypton	680	Ga Al In phosphide
231.4	Ruby ×3	463.4	Krypton	687.1	Krypton
244	Argon ×2	465.8	Argon	694.3	Ruby
248	Krypton fluoride	468.0	Krypton	722.9	Lead
257.2	Argon ×2	472.7	Argon	752.5	Krypton
263	Nd:YLF ×4	476.2	Krypton	799.3	Krypton
266	Nd:YAG ×4	476.5	Argon	852.4	Calcium
275.4	Argon	476.6	Krypton	866.2	Calcium
305.5	Argon	482.5	Krypton	904	Gallium arsenide
308	Xenon chloride	484.7	Krypton	1053	Nd:YLF
312.0	Gold	488.0	Argon	1060	Nd:glass
325	Helium-cadmium	495.6	Xenon	1064	Nd:YAG
333.6	Argon	496.5	Argon	1092.3	Argon
337.1	Nitrogen	501.7	Argon	1130.0	Barium
347.2	Ruby ×2	510.5	Copper	1152.3	Helium-neon
350.7	Krypton	514.5	Argon	1290.0	Manganese
351	Nd:YLF ×3	520.8	Krypton	1300	In Ga As phosphide
351	Xenon fluoride	527	Nd:YLF ×2	1313	Nd:YLF
351.1	Argon	528.7	Argon	1315	Iodine
353	Xenon fluoride	530.9	Krypton	1319	Nd:YAG
355	Nd:YAG ×3	532	Nd:YAG ×2	1500	In Ga As phosphide
356.4	Krypton	534.0	Manganese	1523	Helium-neon
363.8	Argon	539.5	Xenon	2396	Helium-neon
406.7	Krypton	543.5	Helium-neon	2940	Er:YAG
413.1	Krypton	568.2	Krypton	3391	Helium-neon
415.4	Krypton	578.2	Copper	3508	Helium-xenon
428	Nitrogen	595.6	Xenon		

2608–3093 (21 lines)	Hydrogen fluoride	[a] 9160–9840 (45 lines)	Carbon dioxide
3493–4100 (37 lines)	Deuterium fluoride	10400–11040 (58 lines)	Nitrous oxide
5090–6130 (14 lines)	Carbon monoxide	[a] 10070–11020 (50 lines)	Carbon dioxide

a) See Table 2.1, p. 41 for a detailed wavelength listing

Directory of Acronyms and Abbreviations

Bearing in mind the origin of the word 'laser' itself, it is perhaps inevitable that the field of laser applications is associated with a plethora of acronyms and abbreviations. The use of these has, as a matter of deliberate policy, largely been avoided in this book. Nonetheless, abbreviations are common in the current laser literature, and the following list has been selected to assist the reader.

AAS	Atomic absorption spectroscopy
ADP	Ammonium dihydrogen phosphate
AFS	Atomic fluorescence spectroscopy
AM	Amplitude modulation
AO	Acoustic-optic
ASE	Amplified spontaneous emission
AVLIS	Atomic vapour laser isotope separation
BBO	Beta-barium borate
BOXCARS	Box (geometry) coherent anti-Stokes Raman scattering
BW	Bandwidth
CARS	Coherent anti-Stokes Raman scattering
CCD	Charge-coupled device
CD	Circular dichroism
CDR	Circular differential Raman (spectroscopy)
CLSM	Confocal laser scanning microscopy
COIL	Chemical oxygen iodine laser
COMAS	Concentration-modulated absorption spectroscopy
CPF	Conversion (of reactant) per flash
CPM	Colliding-pulse mode-locked (laser)
CRT	Cathode-ray tube
CSRS	Coherent Stokes Raman scattering
CTD	Charge-transfer device
CVD	Chemical vapour deposition
CVL	Copper vapour laser
CW	Continuous-wave
DFDL	Distributed-feedback dye laser
DFT	Discrete Fourier transform
DIAL	Differential absorption lidar

DL	Diffraction limited
DLS	Dynamic light scattering
DOAS	Differential optical absorption spectroscopy
DPL	Diode-pumped laser
DPY	Diode-pumped YAG (laser)
DRIMS	Double-resonance ionisation mass spectrometry
EL	Electroluminescent: Exposure limit
EMR	Electromagnetic radiation
EO	Electro-optic
FDS	Fluorescence dip spectrometry
FEL	Free-electron laser
FFT	Fast Fourier transform
FIR	Far infra-red
FM	Frequency modulation
FPI	Fabry-Perot interferometer
FSR	Free spectral range
FTIR	Fourier transform infra-red
FWHM	Full width at half-maximum
GC	Gas chromatography
GDL	Gas dynamic laser
GTL	Gas transport laser
GVL	Gold vapour laser
HFS	Hyperfine structure
HOE	Holographic optical elements
HORSES	Higher-order Raman spectral excitation studies
HPLC	High-performance liquid chromatography
IC	Internal conversion: Integrated circuit
ICF	Inertial confinement fusion
ICP	Inductively coupled plasma
IM	Intensity modulation
IPL	Iodine photodissociation laser
IR	Infra-red
IRED	Infra-red emitting diode
IRIS	Infra-red interferometric spectrometer
IRMPD	Infra-red multiphoton dissociation
ISC	Intersystem crossing
IVR	Intramolecular vibrational (energy) randomisation
KDP	Potassium dihydrogen phosphate
KLM	Kerr lens mode-locking
KTP	Potassium titanyl phosphate
LADAR	Laser detection and ranging
LAMMA	Laser microprobe mass analysis
LAMMS	Laser microprobe mass spectrometry
LAMS	Laser mass spectrometer

LAS	Laser absorption spectrometer
LASER	Light amplification by the stimulated emission of radiation
LEAFS	Laser-excited atomic fluorescence spectroscopy
LEF	Laser-excited fluorescence
LC	Liquid chromatography
LCP	Laser chemical processing
LCVD	Laser chemical vapour deposition
LDA	Laser Doppler anemometry
LDV	Laser Doppler velocimeter
LEAFS	Laser-excited atomic fluorescence spectroscopy
LED	Light-emitting diode
LEF	Laser-excited fluorescence
LFBR	Laser fusion breeder reactor
LIA	Lock-in amplifier
LIBS	Laser-induced breakdown spectroscopy
LIDAR	Light detection and ranging
LIFS	Laser-induced fluorescence spectroscopy
LIGS	Laser-induced grating spectroscopy
LIMA	Laser-induced mass analysis
LIMS	Laboratory information management system
LIPAS	Laser-induced photoacoustic spectroscopy
LIR	Laser-induced reaction
LIS	Laser isotope separation
LITD	Laser-induced thermal desorption
LMR	Laser magnetic resonance
LOG	Laser optogalvanic (spectroscopy)
LPS	Laser photoacoustic (or photoionisation) spectroscopy
MALDI	Matrix-assisted laser desorption ionisation
MASER	Microwave amplification by the stimulated emission of radiation
MIS	Metal insulator semiconductor
MOS	Metal oxide semiconductor
MPD	Multiphoton dissociation
MPI	Multiphoton ionisation
MPRI	Multiphoton resonance ionisation
MVL	Metal vapour laser
NDT	Non-destructive testing
NIR	Near infra-red
NLO	Nonlinear optics
OA	Optical activity
OCR	Optical character recognition
OEM	Original equipment manufacturer
OGS	Optogalvanic spectroscopy
OMA	Optical multichannel analyser
OODR	Optical-optical double resonance

OPD	Optical path difference
OPO	Optical parametric oscillator
ORD	Optical rotatory dispersion
PARS	Photoacoustic Raman spectroscopy
PAS	Photoacoustic spectroscopy
PC	Photocathode
PCM	Pulse code modulation
PCS	Photon correlation spectroscopy
PDS	Photodischarge spectroscopy
PDT	Photodynamic therapy
PHOPHEX	Photofragment excitation
PM	Polarisation (or phase) modulation
PMT	Photomultiplier tube
PRF	Pulse repetition frequency
PRK	Photorefractive keratectomy
PSD	Phase-sensitive detector
PWM	Pulse width modulation
PZT	Piezoelectric transducer
QE	Quantum efficiency
QED	Quantum electrodynamics
QLS	Quasi-elastic light scattering
RAS	Remote active spectrometer
REMPI	Resonance-enhanced multiphoton ionisation
RGH	Rare gas halide
RIKES	Raman-induced Kerr effect spectroscopy
RIMS	Resonance ionisation mass spectrometry
RIS	Resonance ionisation spectroscopy
RLA	Resonant laser ablation
ROA	Raman optical activity
RRE	Resonance Raman effect
RRS	Resonance Raman scattering
SALI	Surface analysis by laser ionisation
SEP	Stimulated emission pumping
SERS	Surface-enhanced Raman scattering
SFG	Sum-frequency generation
SHG	Second harmonic generation
SIRIS	Sputter-initiated resonant ionisation spectroscopy
SIRS	Spectroscopy by inverse Raman scattering
SLAM	Scanning laser acoustical microscope
SNR	Signal-to-noise ratio
SPM	Self-phase modulation
SRS	Stimulated Raman scattering
SSL	Solid-state laser
TDL	Tunable diode laser

TDS	Time domain spectroscopy
TE	Thermoelectric
TEA	Transversely excited atmospheric (pressure)
TEM	Transverse electromagnetic mode
THG	Third harmonic generation
TIR	Total internal reflection
TLV	Threshold limit value
TOF	Time of flight
TPA	Two-photon absorption
TPD	Two-photon dissociation: Temperature-programmed desorption
TPF	Two-photon fluorescence
TRRS	Time-resolved Raman spectroscopy
UHV	Ultrahigh vacuum
USLS	Ultrafast supercontinuum laser source
UV	Ultraviolet
VDU	Visual display unit
VLD	Visible laser diode
VUV	Vacuum ultraviolet
XUV	Extreme ultraviolet
YAG	Yttrium aluminium garnet
YLF	Yttrium lithium garnet
ZEKE	Zero kinetic energy (photoelectron spectroscopy)

Selected Bibliography

General References

Few textbooks deal comprehensively with lasers in chemistry; most books dealing with the subject are multi-authored conference proceedings. Many of these provide a useful insight into highly topical fields of research, but are not particularly enlightening without a specialised background knowledge; for this reason they have been excluded from this bibliography. The following texts can be recommended for a more general readership:

Ben-Shaul A, Haas Y, Kompa K L, Levine RD (1981). Lasers and chemical change, Springer, Berlin Heidelberg New York
Duley WW (1983) Laser processing and analysis of materials, Plenum, New York
Hecht J (1992) Laser guidebook, 2nd edition, McGraw-Hill, New York

The book by Ben-Shaul et al. is particularly recommended for its very thorough treatment of theory. Although not comprehensive in its coverage, the 'Laser Handbook' series published by North-Holland is also recommended for its detailed and thorough treatment of many topics in this area.

For detailed and up-to-date information on commercially available lasers and laser instrumentation, the reader is referred to the annual Buyers' Guides produced by the journals 'Laser Focus World' and 'Photonics Spectra'.

Laser Theory

Haken H (1985) Light vol 2, North-Holland, Amsterdam
Saleh BEA and Teich MC (1991) Fundamentals of photonics, Wiley, New York
Siegman AE (1986) Lasers, Oxford University Press, Oxford
Wilson J, Hawkes JFB (1989) Optoelectronics: An introduction, 2nd edition, Prentice-Hall, London

Laser Spectroscopy

Andrews DL (ed) (1992) Applied laser spectroscopy: Techniques, instrumentation, and applications, VCH, New York Weinheim Cambridge

Andrews DL and Demidov AA (eds) (1995) An introduction to laser spectroscopy, Plenum, London New York

Brückner V, Feller K-H and Grummt U-W (1990) Applications of time-resolved optical spectroscopy, Elsevier, Amsterdam Oxford New York Tokyo

Demtröder W (1996) Laser spectroscopy, 2nd edition, Springer, Berlin Heidelberg New York

Fleming GR (1986) Chemical applications of ultrafast spectroscopy, Oxford University Press, Oxford

Hieftje GM, Travis JC, Lytle FE (eds) (1981) Lasers in chemical analysis, Humana, Clifton NJ

Kliger DS (ed) (1983) Ultrasensitive laser spectroscopy, Academic, New York

Letokhov VS (ed) (1986) Laser analytical spectrochemistry, Adam Hilger, Bristol

Moenke-Blankenburg L (1989) Laser microanalysis, Wiley, New York

Myers AB and Rizzo TR (1995) Laser techniques in chemistry, Wiley, New York

Laser-Induced Chemistry

Bäuerle D (1996) Laser processing and chemistry, 2nd edition, Springer, Berlin Heidelberg New York

Letokhov VS (1983) Nonlinear laser chemistry, Springer, Berlin Heidelberg New York

Steen WM (1991) Laser material processing, Springer, Berlin Heidelberg New York

Steinfeld JI (ed) (1981) Laser-induced chemical processes, Plenum, New York

Laser Safety

Hughes D (1992) Notes on protection against laser radiation in the laboratory, HHSC, Leeds

Sliney DH (1992) Medical lasers and their safe use, Springer, Berlin Heidelberg New York

Further up-to-date information can be obtained from the Internet, as for example through useful links on the home page of the Laser Institute of America, http://www.creol.ucf.edu/%7Elia/

Answers to Numerical Problems

Chapter 1
1) 0.003 cm^{-1}.
2) 66.7 ns.
3) 1 ms.
4) 44 mrad; 10 μm.
5) (a) 10^7; (b) 100.
6) 0.019.
7) 1.85 × 10^{-30} C m = 0.554 D; 11.6 pm.
8) 0.68 mg (but in air, *radiometric* pressures are several orders of magnitude larger).

Chapter 2
2) 33.8 J; 338 MW.
5) 2.58 cm s^{-1}.
6) $\sqrt{2}$ × (2.6, 3.0) μm = (3.7, 4.2) μm
8) 1.23 μm.

Chapter 3
2) 1559 nm.
4) 358.0 nm.
5) $\lambda(\lambda + \Delta\lambda)/(5\lambda + 4\Delta\lambda)$.
7) 7.61 ps; 6.67 ns.
8) 5235; 1.6 × 10^{15}; 2.5 × 10^{18} W m^{-2}
9) 158 Hz.
10) 2.0 m.

Chapter 4
4) 1342 cm^{-1}; 458.0 nm; I_{AS}/I_S = 0.0026.
7) (a) 9.96 × 10^{-4}; (b) 4.96 × 10^{-7}; (a′) 9.88 × 10^{-3}; (b′) 4.88 × 10^{-5}.

Chapter 5
5) (b) 18.
8) 6.1 × 10^{-3}.

Subject Index

A

Ablation 49, 123, 162
Absorbance 102, 170
Absorption
- coefficient 101
-, - molar 102
Absorption spectroscopy 101
Acousto-optic modulation 67, 73
Actinic range 95
Activation energy, laser-induced
   reaction 179, 182
Active medium 3
- modulation 73
Aerosol particles 87
Alexandrite laser 30
Alignment, optical 17
Allene, CH_2CCH_2 184
Ammonia, NH_3 198, 205
Anemometry 87
Anisotropy
- absorption 186
- fluorescence 125
Anti-Stokes Raman shift 129
Argon fluoride laser 49
Argon ion laser 35
Array detectors 77, 131
Arsine, AsH_3 205
Atmospheric pollution monitoring 17, 45,
   112 *(see also Lidar)*
Atom trapping 201
Atomic beam 201
Atomic fluorescence, laser-induced, *see*
   Fluorescence spectroscopy, laser-induced
Atomic iodine photodissociation laser, *see*
   Iodine laser
Autocorrelation function 84
Autoionisation 170
Aversion response 95

B

Barium borate, beta- 64, 66
Barium enrichment 201

Barium-vapour laser 39
Beam
- divergence 17
- profile 18
- quality 18
- waist 17, 18
Beer-Lambert law 101
Benzene, C_6H_6 154
Bicyclo[2.2.1.]hepta-2,5-diene, C_7H_8 184
Bimolecular reactions, laser-
   enhanced 182
Birefringence 58, 64
Bleaching, *see Saturation*
Blood flow 87
Boltzmann relation 10, 130
Boron trichloride, BCl_3 204, 205
Bose-Einstein distribution 21
Boxcar integrator 80
Breakdown-mode surgery 29
Breakdown spectroscopy, laser-
   induced 123
Bremsstrahlung radiation 54
Brewster angle window 58
Brillouin scattering 85
Bromobenzene, C_6H_5Br 200
Bromomethane, CH_3Br 204
Brownian motion 86
Buffer gases 178
2-Butene, $CH_3CHCHCH_3$ 185

C

Calcite, $CaCO_3$ 58
Calibration, wavelength 114
Cancer therapy 208
Carbon dioxide laser 39
Carbon monoxide laser 43
Carbonyl
- chloride $COCl_2$ 205
- sulphide, OCS 185
Catalysis, *see Heterogeneous catalysis,*
   Homogeneous catalysis
Cataract formation 94

Springer
and the
environment

At Springer we firmly believe that an international science publisher has a special obligation to the environment, and our corporate policies consistently reflect this conviction.

We also expect our business partners – paper mills, printers, packaging manufacturers, etc. – to commit themselves to using materials and production processes that do not harm the environment. The paper in this book is made from low- or no-chlorine pulp and is acid free, in conformance with international standards for paper permanency.

 Springer

Printing: Mercedesdruck, Berlin
Binding: Buchbinderei Lüderitz & Bauer, Berlin